Visual Astronomy

A guide to understanding the night sky

Visual Astronomy

A guide to understanding the night sky

Panos Photinos

Morgan & Claypool Publishers

Rights & Permissions
To obtain permission to re-use copyrighted material from Morgan & Claypool Publishers, please contact info@morganclaypool.com.

ISBN   978-1-6270-5481-2 (ebook)
ISBN   978-1-6270-5480-5 (print)
ISBN   978-1-6270-5712-7 (mobi)

DOI   10.1088/978-1-6270-5481-2

Version: 20150301

IOP Concise Physics
ISSN 2053-2571 (online)
ISSN 2054-7307 (print)

A Morgan & Claypool publication as part of IOP Concise Physics
Published by Morgan & Claypool Publishers, 40 Oak Drive, San Rafael, CA, 94903, USA

IOP Publishing, Temple Circus, Temple Way, Bristol BS1 6HG, UK

To Demetri J, Demetri W, Maria A, Marko W, and Zoe M
And to Shelley A for making it all possible.

Contents

Preface

Successful space missions and technological advancements in communications have, in a way, brought the Cosmos closer to us. The data from various space missions are readily available to the public, and reliable sources offer a wealth of information including coordinates, distances, magnitudes and other properties of astronomical objects. Also, a quick Internet search would show a wealth of images of astronomical objects, near and far. There are beautiful images of the Sun, of galaxies and of nebulae that have inspired many artists, and understandably so. Some of these images are artworks in their own right! In terms of backyard astronomy, technological advances bring us affordable telescopes, with cameras and software that make it very easy to point the telescope to a celestial object and observe or photograph it.

This book is intended for a general audience, with no special background in mathematics or science. The main objective of the book is to provide a concise and self-contained introduction to the basic concepts of observational astronomy. The intent is to help the reader understand the information presented in various resources, and what this information tells us about the motion patterns and appearance of the night sky.

The approach I have adopted is conceptual and mostly qualitative, and the chapters are self-contained as much as possible. Relevant quantitative aspects are included in the appendices for readers who would be interested in numerical specifics. The discussion is more focused on what is accessible to visual observation, primarily members of the Solar System and visible stars. In the case of stars, I have included discussion of absorption spectra, and made quick mention of the wealth of information that can be extracted. I believe it is important to make the reader aware of the method, at least to the level of dispelling the mystery/misconception surrounding these issues.

As is the case with books on astronomy, the distances in all diagrams are not to scale, and the numbers, including properties of nearby stars, are revised frequently. There is little one can do about the scale, other than reminding the reader. Regarding the numerical values, I have made an effort to indicate the approximate nature of the quantities listed.

This book originated from my lecture notes in introductory astronomy classes and practical observation assignments. In preparing the book I had the benefit of listening to students' questions, and learning something about their 'own universe'. In my experience as a teacher, the most encouraging sign is when the student begins to ask questions. I would consider my task successfully accomplished if, for each answer the reader finds in this book, a new question is created in his/her mind.

Acknowledgements

It is a pleasure to acknowledge the help and guidance of Joel Claypool, Ms Jeanine Burke and Ms Jacky Mucklow, the staff at Morgan & Claypool and IOP Publishing, for their expert help. I thank Bruce Alber for sharing the eclipse images (figure 4.8). I thank Robert Black of N Medford High School, and Ms Holly Bensel, St Mary's High School, for helpful discussions. Finally, I thank my wife and star gazing companion Shelley for her continuing support and preparation of the manuscript.

Author biography

Panos Photinos

Panos Photinos is a professor of physics at Southern Oregon University (SOU), where he teaches Introductory Astronomy, Observational Astronomy and Cosmology. Prior to joining SOU in 1989 he held faculty appointments at the Liquid Crystal Institute, Kent, Ohio; St. Francis Xavier, Antigonish, Nova Scotia, Canada; and the University of Pittsburgh, Pennsylvania. He was visiting faculty at the University of Sao Paulo, Brazil; the University of Patras, Greece; Victoria University at Wellington, New Zealand; and the University of Melbourne, Australia. Panos completed his undergraduate degree in physics at the National University of Athens, Greece, and received his doctorate in physics from Kent State University, Kent, Ohio. He started naked-eye observations as a child in the Red Sea, and later upgraded to a pair of Merchant brass binoculars in Alexandria, Egypt, and in his homeland, the island of Ikaria, Greece. Ever since he has visited and stargazed from all five continents, and shared his fascination with the night sky with students of all ages. He lives near Mt Ashland where he enjoys the beautiful skies of Southern Oregon from his backyard with his wife Shelley. This is Panos' first book on Astronomy.

Visual Astronomy
A guide to understanding the night sky
Panos Photinos

Chapter 1

Introduction

1.1 A brief survey of celestial objects and a sense of scale

This chapter provides a brief description of the various objects we see in the sky and some of the characteristics that distinguish them. The celestial objects that we see can be divided into two broad categories: objects within our Solar System and objects beyond our Solar System.

Objects within our Solar System include the Sun, the Moon, the planets and their satellites, asteroids, comets, meteoroids and dust particles. The Sun is by far the largest object and source of energy in the Solar System. Planets and satellites are not self-luminous. They are visible because they reflect sunlight.

Objects beyond our Solar System include stars, nebulas and galaxies. These objects are visible because they emit enormous amounts of energy, up to billions of times more than the Sun. Most stars are actually solar systems of their own, with planets and probably satellites, asteroids, comets, etc. The following is a brief discussion of some characteristics of objects in each category, and how they could be distinguished.

Members of the Solar System are among the brightest in the sky. At their brightest, the naked eye planets (Mercury, Venus, Mars, Jupiter and Saturn) appear brighter than any of the ordinary stars. This is the result of the proximity of the planets to Earth. In addition to the high apparent brightness, the planets seem to follow paths within a narrow zone in the sky, which runs along the familiar constellations of the *zodiac*. Therefore the Sun, Moon and planets will never be seen in the Big Dipper or Southern Cross. Some of the smaller objects, e.g. comets, may stray far beyond the zodiac.

Another way that can help distinguish planets from stars is 'twinkling'. Twinkling, or scintillation, is a result of light travelling through the Earth's atmosphere. Twinkling becomes more noticeable when the apparent (not the actual) size of the object (planet/star) is smaller. The Moon does not twinkle because it has a large apparent size. Stars appear to 'twinkle' more than planets because star distances are so large that they appear as 'points' while planets are much closer to Earth and appear as 'disks'. Scintillation is also more noticeable if the object (planet/star) is low on the horizon because the light received by the observer travels obliquely through the atmosphere, thus the light interacts with a thicker layer of atmosphere.

Due to their orbital motion around the Sun, the distance of planets from Earth and their location relative to the Sun varies in cycles over the year(s). This variation causes noticeable changes in their apparent brightness in the night sky. The variation in brightness is particularly noticeable in the case of Venus and of Mars whose orbits bring them closer to Earth than any other planet. For smaller objects, such as comets, the variation in brightness can be noticeable over a few weeks. Meteors are bright streaks caused by comparatively small sized objects (as small as dust particles in the case of meteor showers) burning off in the Earth's atmosphere. The brightening of meteors is brief and usually spectacular.

The apparent brightness of stars can also change. For some stars the change repeats in cycles of a few days to several months. Contrary to the variation observed in planets, the variation in stars does not involve their distance to Earth. The variation of star brightness can be the result of different processes. For instance, repeated brightening can indicate the existence of a companion star. Also, repeated explosions are known to occur (the so-called *novae*) or terminal explosions marking the end of a star's life (*supernovae*) and other complicated processes that are beyond the scope of this book.

1.2 The size of objects in the Solar System

Within the Solar System, the largest and brightest object is the Sun, with a diameter of about 1.4 million km. The Earth's diameter is approximately 13 000 km. Jupiter is the largest of the planets, with a diameter approximately 11 times larger than the Earth's diameter. Asteroids, comets and meteoroids are much smaller. Ceres, the largest asteroid, has a diameter of about 960 km. Ceres is now considered a dwarf planet like Pluto. Figure 1.1 shows the Sun and the planets to scale. The smaller objects have irregular shapes, while the larger objects assume a more or less spherical shape.

According to the International Astronomical Union (IAU) the distinction between planets and dwarf planets is that planets have enough gravity to attract all nearby debris and clear the space surrounding their orbits. The term debris is used to indicate objects of irregular shape, ranging from dust sized to many kilometers. Contrary to planets, their shapes are very irregular. This is due to their smaller mass and size. Objects with a diameter smaller than approximately 500 km do not have enough gravity to compress them into a roughly spherical shape, thus smaller objects have irregular shapes.

1.3 Objects beyond the Solar System

A few thousand stars are visible to the naked eye under clear sky conditions and moonless nights, and at first sight it may be difficult to discern any patterns. Starting an observation before the sky becomes completely dark (usually about 1 hour after sunset for most locations) allows the observer to see only the brightest of the stars,

Figure 1.1. The Sun and planets. The sizes are to scale, but distances are not to scale. Credit: The International Astronomical Union/Martin Kornmesser, www.iau.org/public/images/detail/iau0603a/.

which serve as a good starting point for recognizing the usual star patterns, the familiar *constellations* and *asterisms*. These patterns can span a comparatively large area in the night sky. For example, the Pleiades or Seven Sisters (a group of stars, or asterism, in the constellation of Taurus) covers an area about two times larger than the apparent size of the full Moon. The constellation of Orion is about 50 times the apparent size of the full Moon.

The vast majority of the objects we see with the naked eye are stars. In terms of actual size, the diameter of ordinary stars is typically of the order of millions to hundreds of millions of kilometers. The use of high power telescopes reveals exotic objects, such as neutron stars, which are very small, approximately 20 km in diameter. Nebulae can be hundreds of trillions of kilometers in size. The Great Orion Nebula, visible to the naked eye, is about 120 trillion km across (about a million times larger than the Sun). Nebulae, such as the Monkey Head Nebula shown in figure 1.2, are huge star nurseries.

While vast, nebulae are still small compared to galaxies. For instance, the *Andromeda Galaxy* which is also visible to the naked eye, is roughly 10 000 times larger than Orion's nebula. By far, the largest star pattern in the sky is the *Milky Way* (MW). All the stars we see belong to our Galaxy, the MW, which appears as a faint band across the sky. It is clearly observable on moonless nights from both the Northern and Southern Hemispheres, running roughly from north–east to south–west around midnight in mid-July. Galileo was the first to report, in his pamphlet *Starry Messenger*, that the MW was actually 'nothing else but a mass of innumerable stars planted together in clusters'. The MW is a large group of stars (over 200 billion) forming a disk structure with spiral arms.

The Sun is an average sized star in the disk of the MW, closer to the edge than the center of the disk. The Sun completes one orbit around the center of the MW in about 250 million years. What we see as a band in figure 1.3 is the disk of the MW.

Figure 1.2. A portion of the Monkey Head Nebula (also known as NGC 2174), a region of active stellar formation. New-born stars are emerging from dust (center-right.) Credit: NASA, ESA and the Hubble Heritage Team (STScI/AURA), http://hubblesite.org/gallery/album/pr2014018a/.

Figure 1.3. The MW. Jupiter is above the MW. Credit: Bruno Gilli/ESO, http://www.eso.org/public/usa/images/milkyway/.

Under a clear sky in the Northern Hemisphere one can see another galaxy, the Great Galaxy in Andromeda (or M31), which is larger than the MW, containing an estimated one trillion stars. M31 is shown in figure 1.4. Because of its immense distance, M31 appears as a fuzzy speck of light. In the Southern Hemisphere one can observe two more galaxies, the so-called *Large and Small Magellanic Clouds* (LMC and SMC, respectively) which are roughly 100 times closer than M31. As their names suggest, the LMC and SMC, shown in figure 1.5, appear as thin clouds not

Credit: T. Rector and B. Wolpa (NOAO/AURA/NSF)

Figure 1.4. A ground based image of the Andromeda Galaxy. Credit: T Rector and B Wolpa (NOAO/AURA/ NSF), http://hubblesite.org/newscenter/archive/releases/2012/04/image/c/.

Figure 1.5. The Large (top middle) and Small (top right) Magellanic Clouds from ESO's La Silla Observatory. Credit: ESO/Y Beletsky, http://www.eso.org/public/usa/images/la-silla-beletsky/.

just specks of light. Although the LMC and SMC appear more extended in the night sky than M31, they are actually much smaller than M31, each containing a few billion stars. They appear larger because they are much closer to Earth.

1.4 Distances of celestial objects

To appreciate the distances involved, we can start with our distance to the Moon, which is approximately 400 000 km (about 250 000 miles). Our distance to the Sun is approximately 400 times larger, which is about 150 million km (about 100 million miles). Compared to our distance to the Moon, the distance to the nearest star (Proxima Centauri) is 100 million times larger (about 40 trillion km). To avoid using huge numbers, astronomers use longer 'yardsticks' which will be introduced next.

Within the Solar System, it is common to use the *astronomical unit* (AU):

$$1 \text{ AU} = 150 \text{ million km.}$$

Beyond the Solar System we can use the *light year* (ly) which is the distance travelled by light in one year:

$$1 \text{ ly} = 9.4 \text{ trillion km.}$$

In terms of AU:

$$1 \text{ ly} = 63\,271 \text{ AU.}$$

The star *Proxima Centauri* is at a distance of 4.2 ly from the Sun, or 266 000 AU.

To obtain a feel for the vast distances between stars, the following comparison may be useful. It takes about 8 min for light from the Sun to reach us on Earth while it takes 4.2 years for light from Proxima Centauri to reach us on Earth. So, comparing our distance to the Sun and our distance to Proxima Centauri is like comparing eight minutes to 4.2 years.

In the above comparison we used 4.2 ly for the distance of the Sun to Proxima Centauri and also for the distance of Earth to Proxima Centauri. This interchange is legitimate because our distance to the Sun is so minute compared to our distance to Proxima Centauri. Obviously, adding or subtracting 8 min (light travel time from Earth to the Sun) to 4.2 years (light travel time to Proxima) does not make much of a difference.

The *parsec* (pc) is a commonly used unit comparable to the light year[1]:

$$1 \text{ pc} = 3.26 \text{ ly.}$$

Although the light year and parsec may appear enormous, multiples of these units are necessary to describe distances beyond our immediate stellar neighborhood in the MW. For example, the distance to M31 is about 2 million ly. This may sound like an enormous distance, but is small by cosmic standards. M31 is one of the nearest neighbors to the MW. The use of high power telescopes reveals objects at distances of billions of light years.

---

[1] For a definition of the parsec see appendices A and B.

Table 1.1. The distances of various objects from Earth.

Object	Distance from Earth
Moon	1.3 light seconds
Sun	8.3 light minutes
Proxima Centauri (nearest star)	4.2 light years
Sirius (brightest star in the sky)	8.6 light years
Pleiades star cluster	approximately 440 light years
Andromeda Galaxy	2.5 million light years
z8_GND_5296 (most distant galaxy at date of writing)	13.1 billion light years

Table 1.1 lists some distances from Earth that illustrate the wide range of magnitudes involved.

A significant difference between objects of the Solar System and stars is that due to the enormous distances involved, stars and other objects beyond the Solar System do not appear to move relative to each other. As a result, within a human lifetime, the star patterns in the night sky (what we generally call asterisms and constellations) appear unchanging. The star patterns that we see today appear similar to what the early Assyrian and Babylonian observers saw.

The fact that the star patterns appear unchanging does not mean that stars do not move. In fact the stars forming each constellation/asterism move independently of each other, typically with speeds of tens or hundreds of kilometers per second (tens or hundreds of thousands of kilometers per hour). To understand the situation it is important to recognize that what we perceive is not the relative distance of the stars, but the direction of our line of sight to the stars. The situation is analogous to comparing a high-flying jet plane traveling at 1000 km per hour and a bird flying at 10 km per hour. The plane appears to move much more slowly than a bird. Obviously the jet travels a larger distance than the bird each second, but our line of sight to the airplane shifts much more slowly because the airplane is much further away.

Because the distances to the stars are so immense the angle formed by our line of sight to a pair of stars does not change noticeably in the span of a human lifetime. Due to the Earth's rotation, the entire sky appears to move. But the relative positions of the stars with respect to each other remain the same and, as a whole, the star pattern appears unchanging.

Planets are much closer to the Earth than the stars, so their locations in relation to each other and to the stars shift noticeably. For the nearest planets (Venus, Mars and Mercury) the shift can be noticeable in a span of a few days. Being much closer than the planets, the shift of the Moon relative to the stars is noticeable over a few hours. In summary, the more distant the object, the smaller is its apparent shift in the sky.

Visual Astronomy
A guide to understanding the night sky
Panos Photinos

Chapter 2

The celestial sphere and apparent motion of the Sun

For distances larger than a few hundred meters, the human visual system is incapable of distinguishing depth. Therefore, in the case of celestial objects, we perceive essentially a two dimensional scene of stars, planets, etc, with no depth, shifting peacefully in the night sky, and together they appear (as the ancient Greeks thought) like a décor. The word *cosmos* derives from the Greek word κοσμος (kosmos) meaning décor. Observing the rapid shift of the Moon, the Sun and the planets among the star patterns, ancient observers concluded that these objects must be much closer to us than the stars and they succeeded in placing these objects in the correct order of distance from the Earth, which is a remarkable achievement for the time.

With no method to gage the distance of stars, the ancient observers generally adopted a model which held all the stars on one or more large spheres. The Earth was at the center of all of these spheres. More spheres were added, one for each of the five planets visible to the naked eye (i.e. Mercury, Venus, Mars, Jupiter and Saturn), the Sun and the Moon. The Earth remained stationary in this model and the spheres revolved around the Earth. This is the *geocentric* model, which prevailed until the 17th century AD.

2.1 The celestial sphere

Today we know that it is the Earth that rotates on its north–south (N–S) axis, and also revolves around the Sun. This is the so-called *heliocentric* model. Yet, in observational astronomy it is practical to use the concept of the *celestial sphere*, which traces back to the geocentric model. The stars are fixed on the celestial sphere and the Earth is at the center of the sphere. As the Earth is spinning on its axis from west to east, it appears to an observer on Earth that the celestial sphere is rotating from east to west.

doi:10.1088/978-1-6270-5481-2ch2   2-1

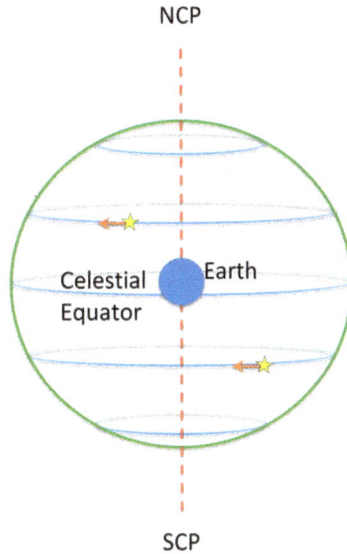

Figure 2.1. The celestial sphere and path of stars.

The direction of the Earth's rotation axis is the line between the *geographic* (which are *not* the same as the magnetic) North and South Poles. Assuming the Earth is a sphere, the Earth's Equator is on the plane that is perpendicular to the N–S axis and divides the sphere in two equal hemispheres. When extended to the celestial sphere, the *direction* of the Earth's N–S axis defines the *north* and *south celestial poles* (NCP and SCP, respectively).

Polaris, also known as the North Star, is a bright star that is very close to the NCP (about 1° off). The SCP at present is marked by *Sigma Octantis*, a star barely visible to the naked eye, which is also about 1° off the pole.

The celestial sphere appears to rotate about the axis defined by the two celestial poles, as shown in figure 2.1. Similarly we can extend the plane of the Earth's Equator onto the celestial sphere and thus define the *celestial equator*. Note that the celestial poles and celestial equator do not refer to actual locations in space. They simply refer to *directions of lines of sight* from Earth.

Any semi-circle from the NCP to the SCP is a *meridian*. For each location, the meridian that crosses the vertical at that location is the *local meridian*. As the celestial sphere rotates from east to west, objects appear to orbit in circles parallel to the equator. This is the *diurnal* (meaning daily) motion. The celestial equator is a great circle because it has the same center as the celestial sphere. Circles parallel to the equator (the small circles, or *parallels*) have their centers on the N–S axis. During their diurnal motion, objects that cross the local meridian, are said to *transit* or *culminate*. For the Sun the transit is what is commonly referred to as noon.

2.2 The annual motion of the Sun

An observation over a few months would show that while the stars stay in the same positions relative to each other and orbit on the same parallel circles on the celestial

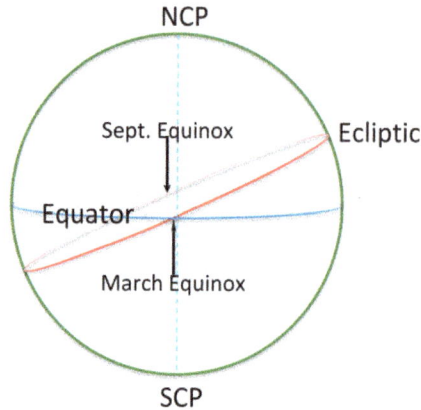

Figure 2.2. The celestial sphere and the ecliptic. The ecliptic crosses the celestial equator at the two equinoxes.

sphere, the positions of objects of the Solar System (Sun, Moon, planets, etc) relative to the stars slowly change and their diurnal motions do not stay on the same parallel circles. Instead they shift with time and the shift follows a periodic pattern, characteristic for each object. For the Sun the repeat cycle is what we know as one year. Over its one-year cycle, the Sun is in the Southern Hemisphere (i.e. below the celestial equator in figure 2.2) of the celestial sphere from about late September to late March, and in the Northern Hemisphere (i.e. above the celestial equator in figure 2.2) from late March to late September. Therefore the Sun is *on* the equator twice a year, when it crosses the celestial equator going from the Southern to the Northern Hemisphere in late March, and also when it crosses the celestial equator going from the Northern to the Southern Hemisphere in late September, as shown in figure 2.2. These two points are known as the *equinoxes* and serve as useful markers on the celestial sphere, as will be discussed in chapter 3. The March equinox is called the vernal (spring) equinox and the September equinox is called the autumnal (fall) equinox. The terms 'March' and 'September' equinox are less ambiguous, but are not generally used.

Figure 2.3 shows the star map in the region of the two equinoxes. The blue line is the *ecliptic*, i.e. the annual apparent path of the Sun. The celestial equator is represented by the black line labeled 0°. North is towards the top, south towards the bottom. The equinoxes occur when the Sun is in the constellations of Pisces (March) and Virgo (September). For historical reasons, the March equinox is referred to as the first point in Aries, although it is in the constellation of Pisces. The March equinox is slowly shifting towards the constellation of Aquarius, as will be discussed in section 2.5. The path of the Sun is from right to left in both maps, which corresponds to a motion from west to east among the stars.

During the March equinox, the path of the Sun crosses from the Southern Hemisphere of the celestial sphere to the Northern Hemisphere. The opposite happens during the September equinox. The labels at the top and bottom of the maps are the right ascension, and will be discussed in chapter 3.

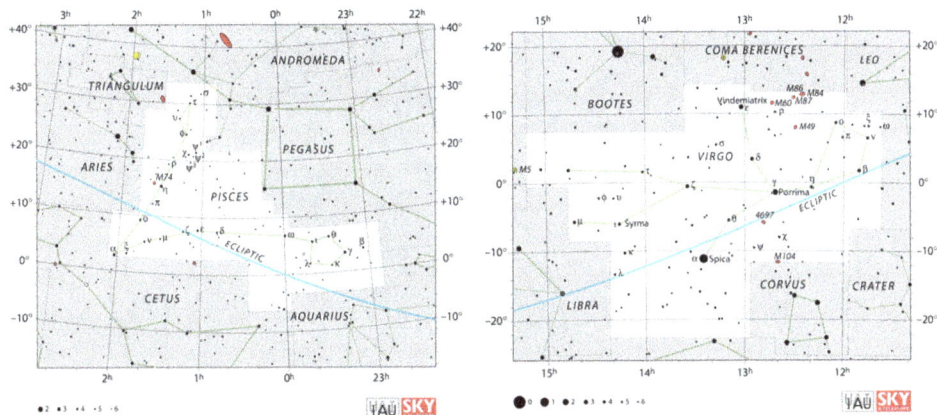

Figure 2.3. A star map of the region around the March equinox (left) and September equinox (right). Credit: International Astronomical Union, and *Sky and Telescope*.

The exact times and dates of the equinoxes are different every year. The time and date listed usually refer to Greenwich time/date, which is different from local time/date[1]. They occur around 20 March and 22 September. The ecliptic, as seen in figure 2.2, is also a great circle on the celestial sphere, but it is not perpendicular to the N–S axis (the equator is). The points where the two major circles (ecliptic and equator) intersect are the equinoxes. The angle between them is about 23.5° and is referred to as the *obliquity* of the ecliptic.

Another important feature of the Sun's diurnal motion is that the time between transits (noon to noon) is not the same from day to day, but changes on a one-year cycle as well. This change reflects the fact the Earth's orbit around the Sun is not a circle but an ellipse. Therefore the Earth's orbital speed is not constant, and it is higher during December and lower in June.

2.3 The Sun's analemma

The combination of the obliquity and Earth's non-uniform orbital speed have an interesting effect on the Sun's position in the sky over the year. If one were to take a series of pictures of the Sun at noon every day (standard time) from the same location, the result would be an asymmetric figure '8' known as the Sun's *analemma*. The shape of the analemma is shown in figure 2.4. The image on the left shows the typical analemma for a mid-latitude location in the Earth's Northern Hemisphere. The image on the right shows the analemma for a mid-latitude location in the Southern Hemisphere.

The asymmetry of the shape (the size of the top and bottom loops are not the same) is primarily the result of Earth's non-uniform orbital speed. The red arrows indicate the progression of time. The figures show that for locations in the Northern Hemisphere the Sun is lower in the sky during December than it is in June. The reverse happens in the Southern Hemisphere. Note also that the cardinal directions

---

[1] See appendix C.

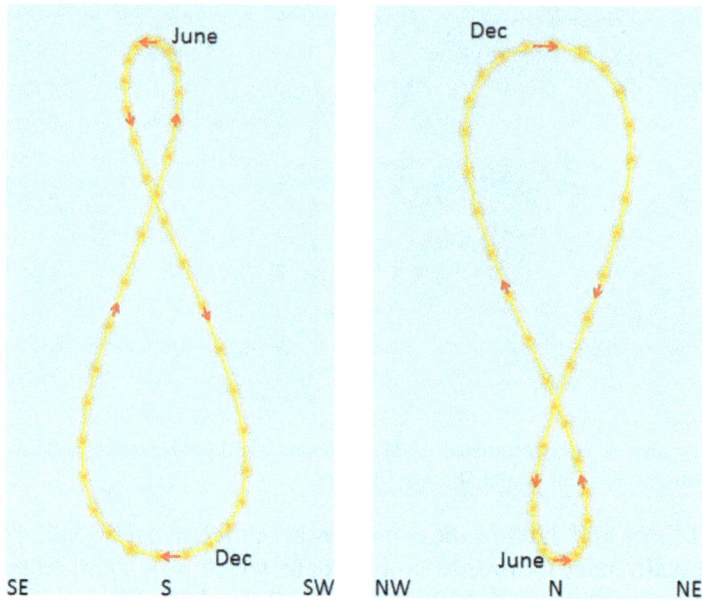

Figure 2.4. The Sun's analemma for mid-latitudes in the Northern Hemisphere (left) and the Southern Hemisphere (right). Note that in the panel on the left the view is towards the south. In the panel on the right the view is towards the north.

are reversed in the two figures. This is due to the fact that for mid-latitudes in the Northern Hemisphere, the Sun at noon is in the southern half of the sky. In the Southern Hemisphere, the Sun at noon is in the northern half of the sky. Note also that the direction of the arrows appears to be reversed: this has to do with the fact that east and west are reversed. The noon position of the Sun shifts from west to east around December and around June in both hemispheres, as indicated by the direction of the arrows at the top and bottom of the figure. The bottom of the analemma curve corresponds to the *winter solstice* and the top of the curve corresponds to the *summer solstice* for each hemisphere.

The solstices occur when the Sun is at its northernmost or southernmost position from the celestial equator (see figure 2.2). Since the Sun's greatest angular separations from the celestial equator are 23.5° and −23.5° (the obliquity of the ecliptic) it follows that the total angular separation between the top and the bottom of the analemma curve equals $2 \times 23.5° = 47°$. The equinoxes are midway between the top and bottom loops in figure 2.4, not the crossing point of the two curves.

2.4 The motion of the sunrise and sunset points

The obliquity of the ecliptic has a direct effect on the sunrise and sunset points. Outside the arctic circles the Sun rises due east and sets due west during the equinoxes. Between the March equinox and the September equinox, the Sun rises north of east and sets north of west. Between the September equinox and the March equinox, the Sun rises south of east and sets south of west. The northernmost sunrise and sunset occur on the June solstice. The southernmost sunrise and sunset occur on the December solstice.

The shift of the sunrise point on the solstices is about 23.5° near the Equator and increases to 90° degrees as we approach the Arctic and Antarctic Circles (latitudes of approximately 66.5° N or S). Within the artic circles, the period between sunrise and sunset can be several months.

2.5 The motion of the celestial poles

Under the influence of gravity, the axis of a spinning top slowly changes direction. Similarly, the gravitational pull of the Sun and the Moon on the Earth's equatorial bulge cause the Earth's rotation axis to change direction in space[2]. This slow motion of the Earth's axis of rotation is called *precession* and has a cycle of about 26 000 years. As a result of precession, the NCP and SCP, which are extensions of the Earth's rotation axis, move gradually among the stars. The NCP was near Thuban (a star in the constellation of Draco) 5000 years ago. Presently, the NCP is near Polaris (a star in the constellation of Ursa Minor) and is shifting slowly towards the constellation of Cepheus. As the celestial equator is perpendicular to the rotation axis of the Earth, the celestial equator is also shifting at the same pace and the same is true for the points where the Sun crosses the celestial equator, i.e. the equinoxes. This shift is called the *precession of the equinoxes* and is shown schematically in figure 2.5.

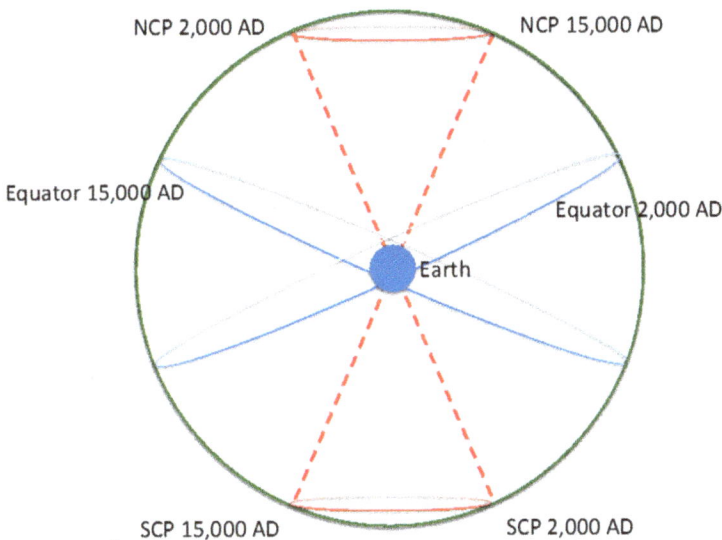

Figure 2.5. The precession of the Earth's axis of rotation. The direction of precession is opposite to the direction of the Earth's rotation. Thus from 2000 to 15 000 AD the NCP will follow the gray half of the circle and the SCP will follow the red half of the circle.

---

[2] The shape of the Earth is not spherical, but oblate. The diameter as measured across the Equator is about 40 km larger than the diameter along the N–S axis.

Visual Astronomy
A guide to understanding the night sky
Panos Photinos

Chapter 3

Coordinate systems

In this chapter we will discuss three commonly used methods to describe the location of objects on the celestial sphere: the equatorial system, the altitude–azimuth system and the ecliptic system.

3.1 The equatorial coordinate system

The equatorial system is the most commonly used coordinate system. This system is essentially an extension of the latitude–longitude system. We already introduced the celestial poles and the celestial equator, as extensions of the Earth's axis and Earth's Equator. The *celestial meridians* or *hour circles* are the great circles from the north to the south celestial poles (NCP and SCP, respectively). The *declination* of a star (δ for short) is defined as the angle between the celestial equator and the line of sight to the star[1]. The declination must be *measured on the meridian* passing through the star. The declination is the equivalent of latitude and is measured in degrees north or south of the celestial equator.

It is common to use positive declinations to indicate north of the celestial equator and negative declinations to indicate south. It is also common to use decimal notation for degrees rather than minutes and seconds. For example, declination of −10.5° indicates 10 degrees and 30 seconds (one half of a degree) south of the celestial equator[2]. The NCP has declination 90° N, or simply 90°, and the declination of the SCP is 90° S, or simply −90°.

To establish an equivalent to the longitude, we need to agree on a prime celestial meridian. By convention, the prime celestial meridian is the meridian through the March (or Vernal) Equinox, also known as the first point in Aries[3]. The equivalent of longitude is the *right ascension* (we will use RA for short). The RA measures the

---

[1] For most objects it does not matter if we use the line of sight form the observer on the Earth's surface, or from the center of the Earth. For the coordinates of the Moon, however, it makes a difference.
[2] See appendix A for conversion to decimal notation.
[3] Because of precession, the first point in Aries is actually in the constellation of Pisces. See figure 2.3.

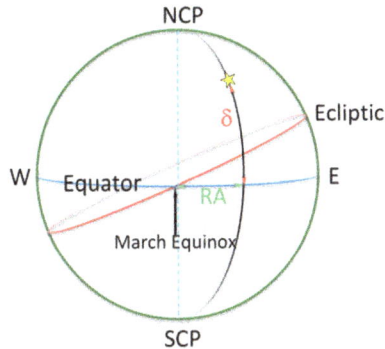

Figure 3.1. RA and declination (δ) of a star.

distance of a star's meridian from the March equinox. The distance is *measured on the celestial equator* from w*est to east*. The RA and declination are shown in figure 3.1.

The RA is an angle and a full rotation corresponds to an angle of 360°. However, as the celestial sphere rotates once every 24 h, there is a correspondence between hours and degrees:

24 h are equivalent to 360°, or
1 hour is equivalent to 15°.

The RA is measured in hours from the March equinox, from west to east. The choice of hours rather than degrees is for convenience as explained below. The RA values are listed in hours, minutes and seconds, for example 3 hours 7 minutes and 20 seconds, or 3 h 7 m 20 s for short[4].

For simplicity we will consider stars that rise and set in a given location. Consider a star of RA 1 h. This means that the star is 1 h (or 15°) east of the March equinox point, since the RA increases in the direction east of the March equinox. As the celestial sphere rotates from east to west, and the RA increases from west to east, it follows that if the March equinox point crosses the meridian at a given instant, then the star with RA 1 h (*any star* with RA 1 h) will cross the meridian 1 hour after the March equinox point. Stars with RA 2 h, will cross the meridian 2 h after the March equinox, and so on.

To illustrate the use of the RA, we consider the two brightest stars in the constellation of Orion

Rigel: RA = 5 h 15 m
Betelgeuse: RA = 5 h 55 m.

Using the RA given above we can compare the transit time of the stars. For example, if Rigel crosses our meridian, then we can predict the transit time of Betelgeuse from the numerical difference between the two RAs:

5 h 55 m − 5 h 15 m = 0 h 40 m.

---

[4] Note that since the rotation is referenced to the stars, the hours are actually sidereal hours, i.e. 1/24 of a sidereal day, as discussed in appendix C.

Rigel has the smaller RA (earlier hour) therefore Rigel transits across the meridian earlier than Betelgeuse by 40 min. For example, if Rigel transits at 22:00 (10 p.m.), then Betelgeuse transits about 22:40.[5] We also note that the calculations above do not involve the declination. In other words, we can use the RA comparisons for any two stars independent of their declination.

An important advantage of the equatorial system is that it is independent of the observer's location and, as such, it is a convenient and widely adopted way of cataloguing stars.

As mentioned in chapter 2, because of the precession of the Earth's axis, the position of the equinoxes among the stars shifts slowly with time. As the RA of a star is measured from the March equinox, it follows that this coordinate changes slowly with time. The shift can be calculated, but for listing purposes it is convenient to refer the RA and declination values as determined at a certain date, or to a certain *epoch*, as it is called. Commonly the listings refer to epoch 1 January 2000, which is indicated by adding the label J2000 to the values of RA and declination.

Figure 3.2 shows a map of the constellation of Scorpius (the scorpion). The vertical lines represent points of equal RA, and the horizontal lines represent points of equal declination. The declination increases from the bottom of the map (which is south) to the top (which is north). The RA increases from the right (which is west) to the left (which is east). This is to be expected because, as mentioned earlier, the RA is measured from west to east. As the celestial sphere rotates from east to west, the stars that are on the right-hand side of the map (smaller RA) cross the meridian first. Note that the grid is not square, because it is a projection of a sphere (the celestial sphere) on the two-dimensional plane of the figure. The declination for the entire constellation is negative, meaning that Scorpius is below the celestial equator.

In the above discussion we considered the equatorial coordinates of stars, however, the coordinates can be used for members of the Solar System, e.g. the Sun, Moon, planets, etc. For members of the Solar System it should be noted that the coordinates change continuously, therefore, these quantities are not listed in tables, constellation maps, etc. The equatorial coordinates for members of the Solar System can be obtained by generating a so-called 'ephemeris' from various sites (see section 3.8).

3.2 The hour angle

Related to the RA is the concept of the *hour angle* (HA for short). The RA is the distance of a star's meridian from the March equinox, measured on the celestial equator from west to east. The hour angle is the distance of a star's meridian (the hour circle of the star, discussed in section 3.1) from the meridian of the observer and is also measured on the celestial equator. The HA is measured from east to west. The HA of a star is zero when a star is on the meridian. It increases as the star moves west of the meridian. The range of the HA is 0 to 24 h. Because the HA of a star is measured with

---

[5] Note that the hours involved here are sidereal hours, therefore, the predictions in terms of civil time are approximate. As explained in appendix C, the difference introduced by using civil time would be 4 minutes at most.

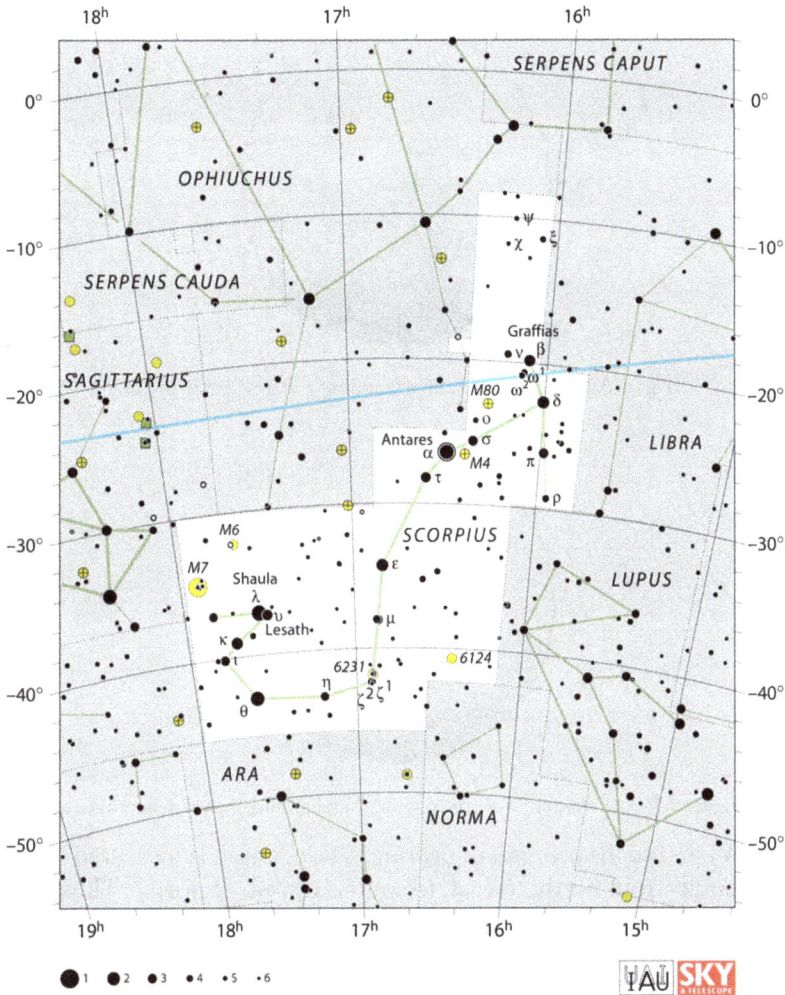

Figure 3.2. A map of the constellation of Scorpius. The blue line is the ecliptic. The 0° line is the celestial equator. Credit: International Astronomical Union, and *Sky and Telescope*.

respect to the local meridian of the observer, the value will be different for observers at different locations. This is an example of a 'local' coordinate, meaning a quantity that does not apply everywhere, and for this reason, it is not listed in tables. The use of the hour angle is discussed further in appendix C.

3.3 The altitude–azimuth coordinate system

Another way to indicate the location of celestial objects is to use the altitude–azimuth system shown in figure 3.3. The altitude measures the angular distance between the horizon and the line of sight to a star. Objects on the horizon line have altitude zero. The highest altitude is 90° and corresponds to the *zenith* (the point directly overhead at the location of the observer) indicated by the letter Z. To indicate the direction

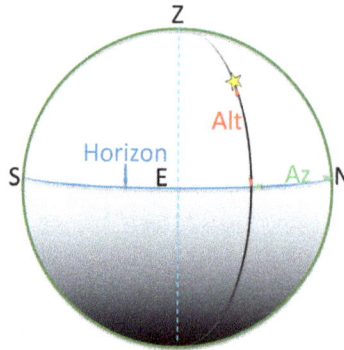

Figure 3.3. The altitude and azimuth of a star. Z is the zenith.

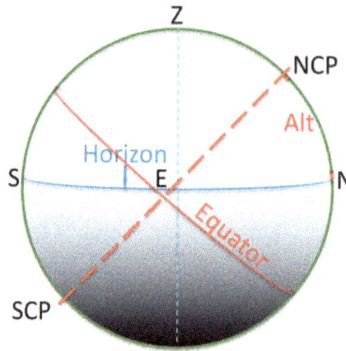

Figure 3.4. The relation between equatorial and altitude–azimuth coordinates.

(N, S, E or W) we use the *azimuth* or *bearing*. The azimuth is measured in degrees on the horizon circle, from north, to east, to south, to west, to north[6]. Thus the azimuth of north is 0°. The azimuth of east is 90°, the azimuth of south is 180° and the azimuth of west is 270°.

The geographical latitude (ϕ) of a location is essentially the declination of the zenith at the particular location. One can use the elementary geometry in figure 3.4 to show that the *altitude of the NCP at any location equals the latitude* at that location. This relation is of particular practical importance and was used by many cultures for navigation.

3.4 The local paths of stars

On the celestial sphere the stars move on circles parallel to the celestial equator (see figure 2.1). The path of stars on the horizon depends only on the latitude of the observer. The following series of figures show the path of stars as they appear from various different latitudes on Earth.

Figure 3.5 shows the path of the stars as seen from at the Earth's poles. The gray shaded area indicates what is below the horizon line.

---

[6] There are different conventions for the starting point and direction, therefore caution is required.

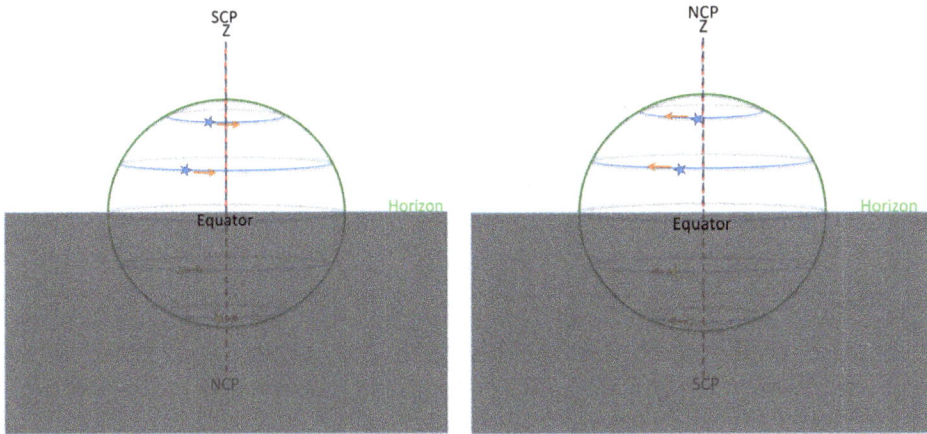

Figure 3.5. The path of stars as seen by observer located at the Earth's South Pole (left) and North Pole (right).

From the poles we note the following:

- The celestial poles (NCP or SCP) coincide with the zenith point as expected because, as discussed in section 3.3, the altitude of the celestial poles is equal to the latitude.
- The celestial equator is right on the horizon.
- The star altitudes above the horizon remain the same. Therefore the stars in the polar regions do not rise and set. The same hemisphere of the celestial sphere is always above the horizon.

As discussed in chapter 2, the Sun spends approximately six months above the celestial equator and six months below the equator. Also, the Sun can be at most 23.5° above or below the celestial equator, which occurs on the solstices. But at the Earth's poles, the celestial equator is right on the horizon, meaning that at the Earth's poles the Sun is approximately six months above the horizon and approximately six months below the horizon[7].

Stars that remain always above the horizon are called *circumpolar stars*. All the stars observed from the Earth's poles are circumpolar.

Figure 3.6 shows the path of the stars as seen from the Earth's Equator.

From the Earth's Equator we note the following:

- The NCP and SCP are right on the horizon, as expected, since the latitude of the equator is 0°.
- The celestial equator arches from east to west and through the zenith. Therefore, the Sun at noon is within 23.5° of the vertical, at most.
- *All* the stars in the celestial sphere rise and set, meaning that over the year the observer sees *all* the stars in the celestial sphere.

---

[7] The solar disk is not a point, but has an extent of 0.5 degrees. It is the center of solar disk that is six months above and six months below the horizon.

- The stars rise vertically to the horizon in the east and set vertically in the western horizon.

Figure 3.7 shows the path of the stars as seen from two mid-latitude locations in the Northern and Southern Hemispheres.

From mid-latitudes we note the following:

- The altitude of the NCP is equal to the latitude of the location.
- In the Northern Hemisphere the celestial equator is in the southern horizon. In the Southern Hemisphere the celestial equator is in the northern horizon.
- Some stars are circumpolar, others rise and set, and some stars never rise above the horizon. Which stars rise and set and which are not visible at all can be determined by comparing the declination of the star and the latitude at the observer's location, as will be discussed next.

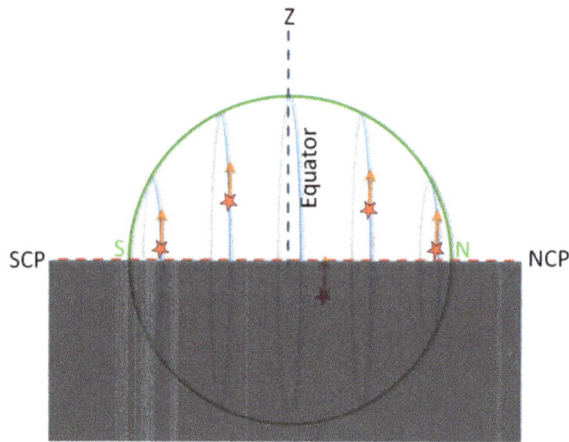

Figure 3.6. The path of the stars as seen by an observer located at the Earth's Equator.

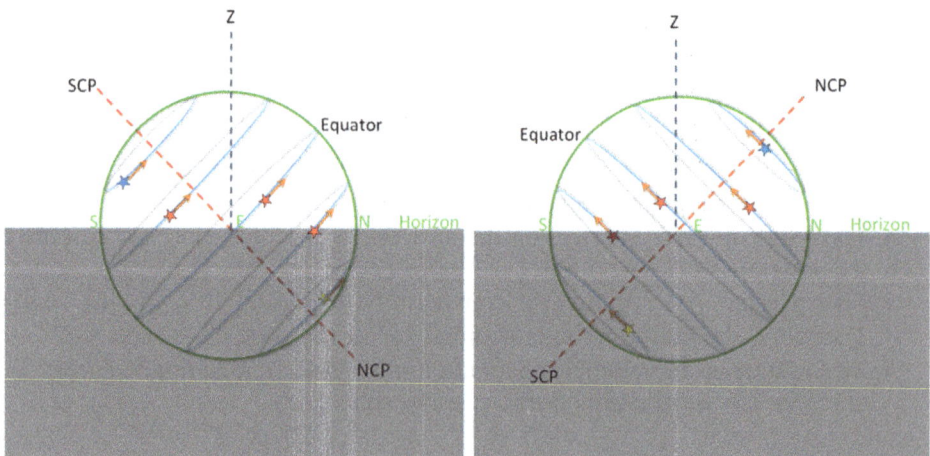

Figure 3.7. The path of stars as seen by observers located at mid-latitudes in the Southern Hemisphere (left) and in the Northern Hemisphere (right).

3.5 Which stars can be seen from a given location?

To find which stars are visible from a given location we need to consider the latitude of the location and the declination of the star. Here it is important to note that locations south of the Earth's Equator have negative latitudes, and stars south of the celestial equator have negative declinations.

For a location of latitude ϕ in the *Northern* Hemisphere:

1. Stars that have declination smaller than $\phi - 90°$ will remain under the horizon at all times (i.e. these stars never rise in that location).
2. Stars that have declination larger than $90° - \phi$ be will be above the horizon at all times (i.e. these star never set in that location). Stars in this category are *circumpolar* in that location.
3. The intermediate category includes the stars that rise and set in a particular location. Stars in this category have declination larger than $\phi - 90°$ but smaller than $90° - \phi$.

For a location of latitude ϕ in the *Southern* Hemisphere:

1. Stars that have declination larger than $\phi + 90°$ will remain under the horizon at all times (i.e. these stars never rise in that location).
2. Stars that have declination smaller than $-\phi - 90°$ be will be above the horizon at all times (i.e. these star never set in that location). Stars in this category are *circumpolar* in that location.
3. The intermediate category includes the stars that rise and set in a particular location. Stars in this category have declination larger than $-\phi - 90°$ but smaller than $\phi + 90°$.

Table 3.1 lists 12 bright stars covering a range of RAs and declinations. The headings of the six columns on the right indicate locations of various latitudes

Table 3.1. Examples of stars that are circumpolar (C), rise and set (RS), or never rise (NR) from various latitudes. Data from http://simbad.u-strasbg.fr/simbad/sim-fid.

Star	RA	Dec.	$\phi = 75°$	$\phi = 50°$	$\phi = 25°$	$\phi = -25°$	$\phi = -50°$	$\phi = -75°$
β Carinae	9 h 13 m	−69.7°	NR	NR	NR	C	C	C
Acrux	12 h 27 m	−63.1°	NR	NR	RS	RS	C	C
Fomalhaut	22 h 58 m	−29.6°	NR	RS	RS	RS	RS	C
Antares	16 h 29 m	−26.4°	NR	RS	RS	RS	RS	C
Sirius	6 h 45 m	−16.7°	NR	RS	RS	RS	RS	RS
Procyon	7 h 39 m	5.2°	RS	RS	RS	RS	RS	RS
Aldebaran	4 h 36 m	16.5°	C	RS	RS	RS	RS	RS
Arcturus	14 h 16 m	19.2°	C	RS	RS	RS	RS	RS
Alpheratz	0 h 8 m	29.1°	C	RS	RS	RS	RS	NR
Vega	18 h 37 m	38.8°	C	RS	RS	RS	RS	NR
Deneb	20 h 41 m	45.3°	C	C	RS	RS	NR	NR
Polaris	2 h 32 m	89.3°	C	C	C	NR	NR	NR

on Earth and which stars would be visible for the particular locations. For example from locations of $\phi = 50°$: β Carinae and Acrux never rise (NR); Fomalhaut through Vega rise and set (RS) in that location, while Deneb and Polaris are circumpolar (C).

Note that the classification is not dependent on the observer's longitude, only the latitude. Also, the classification is not dependent on the RA of the star. The RA determines the time of the year that a given star would be in the night sky, as will be discussed below.

3.6 When can a star be seen in the night sky?

A star's equatorial coordinates, declination and RA are the same day after day. For a given location on Earth, the same stars cross the sky day after day. In contrast, the declination and RA of the Sun changes over a cycle of one year. It follows that the stars that have the same RA as the Sun at a given time of the year will cross the local meridian at the same time as the Sun. Therefore these stars will not be visible in the night sky at that time of the year. Obviously the stars that are circumpolar in a particular location will be up every night. For stars that are seen to rise and set in a given location, we need to consider the RA of the star and the RA of the Sun in order to find the time of the year when the star would be in the night sky (as opposed to being up during daytime)[8]. Table 3.2 lists the Sun's approximate RA for a 12 month cycle.

One way to use the RA is to select a time of the year and ask which stars will cross the meridian at midnight (or any other time of the day). The procedure is as follows:

For a star to cross the meridian at midnight (i.e. 12 h after the Sun crosses the meridian) the star's RA must be equal to the Sun's RA + 12 h. For a star to cross

Table 3.2. Approximate RA of the Sun over a 12 month period.

Date	RA (h)
20 Mar	0
20 Apr	2
20 May	4
20 Jun	6
20 Jul	8
20 Aug	10
20 Sep	12
20 Oct	14
20 Nov	16
20 Dec	18
20 Jan	20
20 Feb	22

---

[8] Here we will assume standard time (as opposed to daylight savings time). See appendix C.

the meridian at 21:00 (9 p.m., i.e. 9 h after the Sun crosses the meridian) the star's RA must be equal to the Sun's RA + 9 h, and so on. For simplicity in the following calculations we will use RA in hours and ignore the minutes.

Example. In late March to early April, the Sun's RA = 0, as seen from table 3.2. Which stars will be crossing the meridian at midnight in late March to early April? Stars that have RA = 0 + 12 = 12 h. From the stars listed in table 3.1 we find that the star Acrux has RA = 12 h, therefore Acrux will be up in the sky at midnight.

Example. Which stars will be crossing the meridian at 21:00 (9 p.m.) in late March to early April? All the stars that have RA = 0 + 9 = 9 h. From the stars listed in table 3.1 we find that the star β Carinae has RA = 9 h, therefore β Carinae will be crossing the meridian at 21:00 (9 p.m.).

Example. In late May to early June the Sun's RA = 4 h, as seen from table 3.2. All stars that have RA = 4 + 12 = 16 h will cross the meridian at midnight. Form table 3.1 we find that Antares will be crossing at midnight.

Another way to use the RA is to ask what time a given star crosses the meridian at a given time of the year, for example late April to early May. To find the time, we subtract the Sun's RA at that time of the year, from the RA of the star, and add the result to 12 noon.

Example. In late April to early May the Sun's RA = 2. The star β Carinae has RA = 9 h. The difference between RAs is 9 − 2 = 7 h. Adding the difference to 12 noon, we find 12 + 7 = 19 h, i.e. the star β Carinae will cross the meridian at 19 h (19:00) by the 24 hour clock or 7 p.m. by the 12 hour clock.

If the difference is negative, we add the negative number to 12 noon.

Example. In August to September, the Sun's RA = 10 h. The star Alpheratz has RA = 0 h. The difference of RAs is 0 − 10 = −10 h. Adding −10 h to 12 noon we have 12 − 10 = 2 h, i.e. the star Alpheratz will cross the meridian at 2 h (02:00) by the 24 hour clock, or 2 a.m.

Note that the calculations above do not involve the latitude, or longitude of the observer. As long as the star is visible from a location, the calculated times apply.

In planning an observation, one should also consider the sunset time. In particular, for dark sky, one should wait for the Sun to go well below the horizon. In astronomy, the night sky is dark enough at the end of the so-called *astronomical twilight*, which means when the Sun goes 18° below the horizon. Near the Equator, the astronomical twilight comes about 40 min after sunset, because the path of the stars (and the Sun) is close to vertical (see figures 3.5–3.7). As we move away from the Equator, the astronomical twilight ends 1 hour or more after sunset. Near the polar regions, the astronomical twilight may not end for several months.

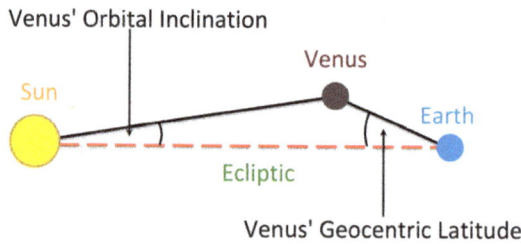

Figure 3.8. The geocentric ecliptic latitude of Venus is larger than the inclination of Venus' orbit when Venus is close to Earth. The figure is not to scale.

3.7 The ecliptic coordinate system

The ecliptic coordinate system uses longitude and latitude. The latitude is measured in degrees north or south of the *ecliptic*. The longitude is measured in degrees *along the ecliptic*. The zero longitude is the March equinox and the angle is measured from *west to east*, i.e. in the same direction as the RA in the equatorial system. The center of the system can be the Earth (*geocentric*) or the Sun (*heliocentric*). There is no special designation for the poles of this system. The poles are defined by the perpendicular to the plane of the ecliptic. As the ecliptic is inclined by 23.5° to the celestial equator, the poles of the ecliptic system are separated by an angle of 23.5° from the celestial poles. This system is useful for describing the location of objects in the Solar System: the Moon, planets, etc.

The orbits of the Moon and the planets are very near the ecliptic, therefore their ecliptic latitude is small. For example, the Moon's orbit is inclined by about 5° to the ecliptic, i.e. the Moon can be 5° above or below the ecliptic. Therefore, the geocentric latitude changes between −5° and +5°. The inclination of the planetary orbits with respect to the ecliptic are measured from the Sun. For example, the orbit of Venus is inclined by about 3° to the ecliptic. However, the *geocentric ecliptic latitude* of Venus can be as much as + or −8° when Venus is at its closest distance to Earth, as indicated in figure 3.8.

The geocentric ecliptic longitude of the Moon increases continuously from 0° to 360° in about 27.5 days. The ecliptic longitude of the planets changes in a more complicated way, as discussed in chapter 5.

3.8 Resources

The *coordinates of stars* are available from http://simbad.u-strasbg.fr/simbad/sim-fid and http://nssdc.gsfc.nasa.gov/.

The *coordinates of planets* can be generated and downloaded in the form of an 'ephemeris' from http://ssd.jpl.nasa.gov/horizons.cgi#top.

IOP Concise Physics

Visual Astronomy
A guide to understanding the night sky
Panos Photinos

Chapter 4

The motion and phases of the Moon

To understand the changes in the appearance and the path of the Moon, it is important to keep in mind that there are three cycles involved:

a. The Earth *rotates* (spins) on its axis approximately every 24 h.
b. The Moon *revolves* (orbits) around the Earth approximately every 28 days in the same direction as the Earth's spin.
c. The Earth *revolves* around the Sun, completing one orbit approximately every 365.25 days, in the same direction as the Earth's spin.

With every rotation of the Earth (about 24 h) the Moon advances in the same direction by 1/28 of a day (which is about 50 min) therefore, the Moon crosses the observer's meridian every 24 h and 50 min on average. The same applies on average to the Moon rise and set times: they occur 50 min later every day. This number varies by a few minutes from day to day, and depends on the observer's latitude as well. For locations near the poles, the delay within a month can range from about 10 min to 90 min or more. For mid-latitudes, the delay within a month can range from about 35 min to 65 min. Near the Equator the variation is smaller. This variation reflects the fact that the orbit of the Moon is not a circle.

The Moon's orbit is an ellipse, and the (center-to-center) distance from Earth can vary considerably, from about 357 000 km to 407 000 km. As a result, the Moon moves faster when it is closer to Earth, therefore there is less delay in the rise (or set) time from day to day[1]. As we move closer to the poles the variation in the delay time is amplified by the low inclination of the apparent path with respect to the local horizon[2].

The Moon is visible because it reflects sunlight and its appearance depends on its location with reference to the Sun (as seen from Earth). The result is repeating

---

[1] See chapter 5.
[2] See figures 3.5–3.7.

cycles in the appearance of the Moon. Each cycle is a sequence of *phases* (full disk, half disk, etc). Every time the Moon returns to the same relative position with respect to the Sun (as seen from Earth) a new cycle starts and we speak of the *synodic* period. The lunar synodic period is what we know as a *lunar month*. The starting point of the cycle is arbitrary; in the case of the Moon the choice is usually from the very thin sliver (the 'new' Moon phase, as explained below). The location of the Moon can also be referenced to the stars. The time required to complete a cycle and return to the same position on the star map is called the *sidereal period* of the Moon.

4.1 The phases of the Moon

At any given time, one hemisphere of the Moon is illuminated by sunlight, and depending on the Earth's position at the time, we may see various fractions of the illuminated half, hence the phases of the Moon. The illuminated half of the Moon is always on the side facing the Sun. What part (if any) of the lit side will be visible depends on the location of the observer.

Figure 4.1 shows the configurations for the lunar phases. The figure is a top view (from the north side of the Solar System). The quarter phases occur when the illumination is at 90° to our line of sight to the Moon and we see a half-lit disk. For the phases on the left side of the figure, we see more of the unlit half of the Moon and the disk is less than half-lit from an Earth perspective. These are the *crescent* phases. For the phases on the right half of the figure, the Earth (observer) is on the same side as the Sun and the disk is more than half-lit from an Earth perspective. These are the *gibbous* phases. At *full Moon* we see 100% of the illuminated half and at the *new Moon* 0% (no Moon at all). The term *waxing* is used to indicate the increase from the new to the full Moon. The term *waning* describes the decrease from full to new Moon. Therefore, the phases between the new Moon and first quarter are *waxing crescent*

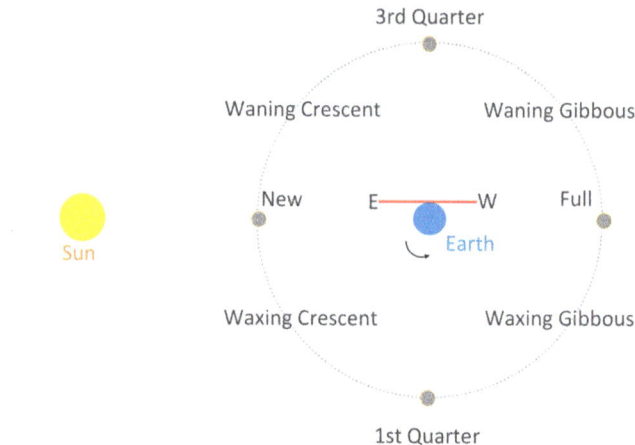

Figure 4.1. The phases of the Moon. The red line indicates the observer's horizon. The curved arrow indicates the direction of the Earth's rotation. At the location shown, the Sun is about to rise. If the Moon were new, then it would be rising at this time. If the Moon were in the 3rd quarter phase, it would be crossing the local meridian of the observer at this time, etc.

phases. From the *first quarter* to the full Moon we have the *waxing gibbous* phases. From the full Moon to the *third quarter* we have the *waning gibbous*. From the third quarter to the new Moon we have the *waning crescents*.

During a synodic cycle the moonrise and moonset times follow a very regular pattern in relation to sunrise and sunset. In figure 4.1 north is above the plane of the image, therefore we are looking south. If the observer is looking south, then east is on the left hand side and west is on the right. The Moon orbits the Earth in the counterclockwise direction (west to east) once every four weeks and the Earth is spinning counterclockwise (west to east) once every 24 h. Therefore in examining the Moon's rise and set times we can assume that the Moon is stationary and consider only the rotation of the Earth.

In the new Moon phase, the Sun and Moon are in the same direction in the sky. Therefore, they rise and set at about the same time. During the full Moon phase, the Sun and Moon are in opposite directions, therefore they rise 12 h apart and they set 12 h apart, i.e. the full Moon rises approximately at sunset and sets at sunrise. When the Moon is in the 1st quarter position in the diagram, due to the counterclockwise rotation of the Earth (west to east), an observer will see the Sun rise first and it will take another quarter-turn for the Earth (i.e. about 6 h) for the Moon to rise above the observer's horizon. Therefore, the 1st quarter Moon rises 6 h after sunrise (i.e. about noon). The same applies to the setting time: the 1st quarter Moon sets approximately 6 h after sunset, i.e. about midnight. The situation is reverse in the 3rd quarter phase. The observer sees the Moon rise and it will take another quarter-turn of the Earth (i.e. about 6 h) for the Sun to rise. Therefore the 3rd quarter Moon rises about 6 h before sunrise (about midnight) and sets approximately 6 h before sunset (i.e. about noon).

Note that the term 'quarter' refers to ¼ of the lunar cycle, not the fraction of the visible surface. Also note that the time between quarters is not the same. The period from the 3rd quarter to the new Moon is not the same as the new Moon to the 1st quarter, etc.

4.2 The diurnal motion of the Moon

As the Moon completes one orbit in about 4 weeks, it will take about 1 week for the Moon to move from the location marked 'new' to the 1st quarter. Another week to move from there to the location marked 'full' etc. A 360° rotation is completed in about 4 weeks, which amounts to 13° per day, or 0.5° per hour. The angular size of the Moon[3] spans about 0.5° which means that from an Earth perspective, the Moon shifts with respect to the star patterns by one Moon diameter every hour. The shift with respect to the star patterns is from west to east. As the entire celestial sphere (Moon, stars, etc) moves from east to west, to an observer on Earth, it appears that the stars are gaining on the Moon by a Moon diameter every hour. As they gain on the Moon, the stars may disappear behind the Moon's unlit side. For example, in early evening during the 1st quarter we can occasionally observe some of the bright

---

[3] See appendix A.

stars (Aldebaran, Antares, etc) disappear behind the unlit east limb (edge) of the Moon. This phenomenon is called *occultation*. The occultation occurs quite abruptly, with no twinkling[4], because the Moon lacks an atmosphere.

4.3 East and west elongation

It is common to use the concept of *elongation* to describe the location of objects relative to the Sun for an Earth perspective[5]. Elongation is the angle between the line of sight from Earth to the Sun and the line of sight from Earth to the Moon. The elongation is measured from the Sun to the east and from the Sun to the west. With reference to figure 4.1, we see that at the new Moon phase, the elongation of the Moon is zero. The 1st quarter phase occurs when the Moon's elongation is about 90° east (of the Sun) and the 3rd quarter occurs when the Moon's elongation is about 90° west (of the Sun). The full Moon phase occurs at elongation of 180°. It is also common to use the *phase* angle, which is the angle between the line of sight from the Moon to the Earth and the line of sight from the Moon to the Sun. In figure 4.1, the phase angle for the full Moon is 0°, for the 1st and 3rd quarter it is 90° and for the new Moon 180°.

4.4 Features of the Moon

The brightness of the lunar disk is not uniform, instead it shows grey patches, the *maria* (plural of *mare*, meaning sea) surrounded by the brighter *terrae* (plural of *terra*, meaning land or highland). The terrae are more rugged and have several mountain chains. Of course there are no (and apparently, there never were) seas on the Moon. The relatively short distance to the Moon allows observation of considerable detail, particularly when the Moon is away from the full Moon phase. During the full Moon phase, the sunlight falls vertically on the center of the Moon's visible surface and, as a result, the shadows of the mountains are very short. Away from the full Moon phase and especially in the parts near the terminator (i.e. the boundary between the bright and dark portions of the disk), the sunlight falls at a very oblique angle, which makes the shadows very long and thus adds contrast. The first feature to become visible after the new Moon phase is towards the north of the crescent, and it is the *Mare Crisium* (Sea of Crises). It is also the first to disappear after the full Moon. Figure 4.2 shows the waxing crescent phase and Mare Crisium.

Note that the view of figure 4.2 is for northern latitudes. The illuminated side is pointing towards the west (where the Sun is). From locations in the Southern Hemisphere, the figure is rotated by 180° so that the illuminated side of the waxing crescent is on the left and Mare Crisium is on the lower left. The illuminated side still points towards the west (where the Sun is) as the observer now is looking towards the northern horizon. At times the tips of the rising or setting crescent can point almost directly upward from the horizon, and the lit side appears like a 'smile' as shown in figure 4.3. This is the *wet* or *Cheshire* Moon and occurs more often in locations near

---

[4] See chapter 1.
[5] The elongation is equivalent to the ecliptic longitude. See chapter 3.

Figure 4.2. The waxing crescent Moon. Mare Crisium is on the northern hemisphere of the Moon (top half of the figure). Credit: ESO and Andy Strappazzon, http://www.eso.org/public/images/potw1129a/.

Figure 4.3. Crescent Moon low in the sky. Credit: ESO/C Malin, http://www.eso.org/public/images/malin_2781/.

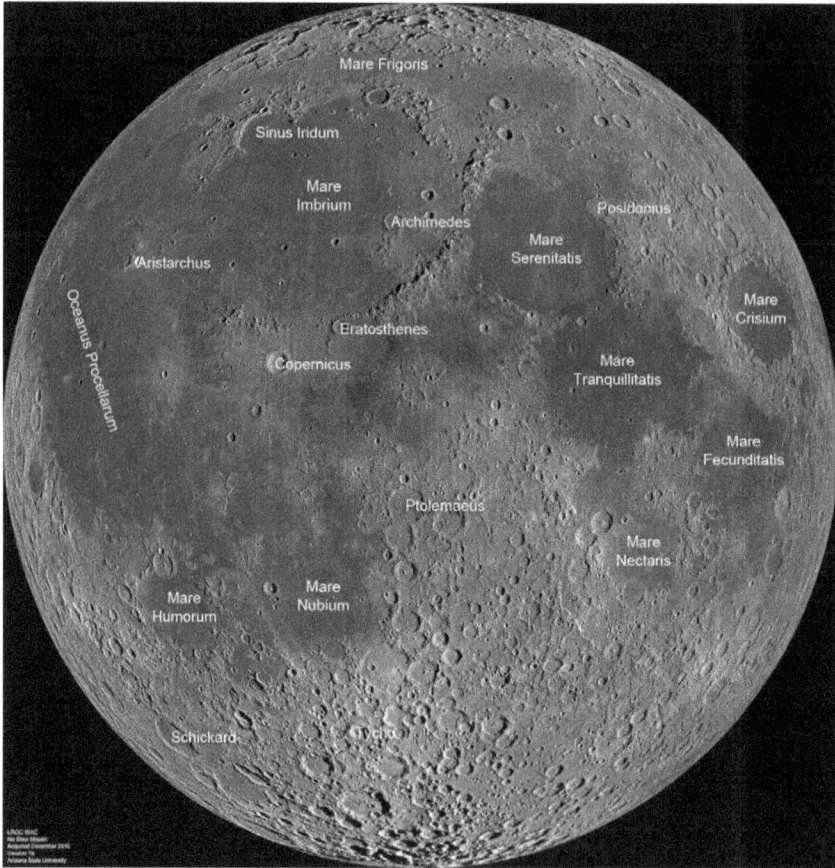

Figure 4.4. The craters and seas (maria) of the lunar topography. Image credit: NASA/GSFC/Arizona State University, http://photojournal.jpl.nasa.gov/catalog/PIA14011.

the Equator, where the ecliptic (and the path of the Moon) is not far from the vertical[6].

An interesting phenomenon occurs in the first few days after the new Moon. Light reflected from the Earth's atmosphere hits the Moon and, as a result, we see a faint image of the remainder of the disk. This is also shown in figure 4.3 and it is called the *old Moon in the arms of the new Moon.*

Figure 4.4 shows the maria on the full Moon. The maria (name preceded by 'Mare') and the 'Ocean' on the left are visible to the naked eye and can be used to navigate through the lunar topography. Names not preceded by 'Mare' are craters. The larger craters are also visible to the naked eye, although a pair of binoculars would convincingly show them as craters. Note that the figure is a composite. As discussed earlier, during the full Moon phase the contrast is very low.

---

[6] See figure 3.4.

4.5 The Moon's orbital path

The path of the Earth is on the ecliptic and the path of the Moon is very close to the ecliptic. The Moon's orbit is inclined by about 5° with respect to the ecliptic. As a result, the Moon's location cycles from 5° above to 5° below the ecliptic and back every 4 weeks. It is worth noting that we see only one side of the Moon, the so-called *near side* of the Moon[7]. The reason we do not see the *far side* of the Moon *is not* because it is dark and also it *is not* because the Moon does not spin. *The reason is that the Moon spins around its axis exactly in sync with the orbital period.*

4.6 Eclipses

Note that the line of sight from the Earth to the Sun is always on the ecliptic, because the ecliptic marks the path of the Sun on the celestial sphere. The Moon's path crosses the ecliptic at two points called the *nodes*. Note that we are speaking of apparent paths, i.e. lines of sight from Earth: the actual orbit of the Moon does not cross the orbit of the Sun. The solar and lunar eclipses can occur only when the Earth, the Sun and the Moon line up. This happens only when the Moon is near the nodes[8]. From figure 4.1 we see that the Moon and Sun are close to alignment with the Earth only during the new or full Moon phase. Therefore the eclipses can occur only during the new or full phase. *Solar eclipses can only occur in the new Moon phase*, i.e. when the Moon is between the Earth and Sun, and therefore can block the sunlight from reaching Earth. *Lunar eclipses can only occur in the full Moon phase*, i.e. when the Earth is between the Sun and the Moon, and therefore can block the sunlight from reaching the Moon.

4.6.1 Lunar eclipses

During a lunar eclipse, the Earth's shadow sweeps the Moon starting from the east limb and proceeding towards the west limb, as shown in figure 4.5.

The shadow of the Earth is divided into two parts: the umbra, which is the region where the sunlight is completely blocked, and the penumbra, where the light is partially blocked. The umbra and penumbra are shown in figure 4.6. Depending on which part of the shadow is cast on the Moon, we have a *total eclipse* (if the entire Moon crosses the umbra) or a *partial eclipse* (if only part of the Moon crosses the umbra).

If the Moon crosses only the penumbra, we speak of a *penumbral eclipse*. Typically, a total eclipse will last approximately 6 h, with the *totality* (Moon entirely in the umbra) lasting for about 2 h. It is interesting that during the totality, the Moon becomes reddish in color (what the popular media call the *blood Moon*). The reddish color is caused by the interaction of sunlight with the Earth's atmosphere. First, the round shape of the Earth's atmosphere acts as a lens that bends part of the sunlight towards the Moon. The bending of the light by the Earth's atmosphere is called

---

[7] Actually we see slightly over 50% of the surface, as the Moon's orbit carries it by 5° above and below the ecliptic, and our line of sight takes us slightly below and slightly above the Moon, respectively.

[8] The line of nodes actually moves, completing a turn every 18.6 years. This motion causes a repeat cycle in the eclipse conditions and was used from ancient times to predict eclipses.

Figure 4.5. The lunar eclipse of 21 February 2008. The time sequence proceeds from left to right. The entire duration is approximately 3.5 h. Credit: ESO, http://www.eso.org/public/images/annlocal08001b/.

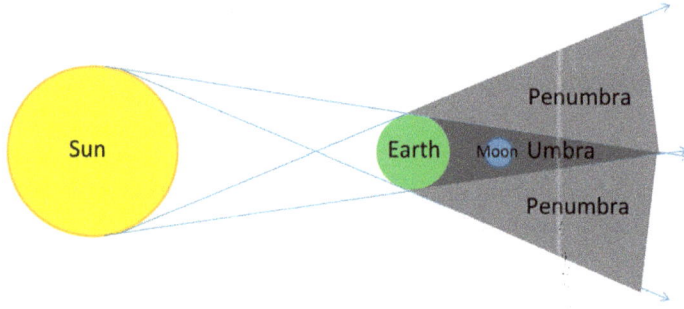

Figure 4.6. The geometry of a lunar eclipse. The figure is not to scale.

atmospheric refraction. Second, when passing through the Earth's atmosphere, the blue–green part of the sunlight is filtered out (like in sunsets) and what remains is light of reddish color that is bent towards the Moon.

Typically, a lunar eclipse would be visible from most locations on Earth, if the Moon happens to be above the horizon at such locations during any part of the eclipse. For example, if the eclipse lasts about 6 h, and assuming that the Moon is above the horizon approximately 12 h, the eclipse can be visible from approximately $(6 + 12)/24 = 0.75$ or 75% of the Earth's locations that can see the Moon on that day. On and near the poles, the Sun/Moon can be below the horizon for extended periods of time therefore eclipses may not be visible.

An interesting phenomenon can be observed in some locations on Earth. If the eclipse is observable near sunrise or sunset in certain locations, because of atmospheric refraction, both the eclipsed full Moon and the Sun may appear to be above the horizon in these locations. The phenomenon is called a *horizontal eclipse*.

4.6.2 Solar eclipses

Solar eclipses occur when the Moon's shadow sweeps over the Earth's surface. Solar eclipses can occur only in the new Moon phase. Figure 4.7 shows the geometry of the solar eclipse.

As the Moon is smaller than the Earth, the Moon's shadow is smaller than the Earth. The dimensions of the shadow depend on the Earth–Moon distance and the angle at which the shadow is cast. Typically the umbra of the Moon' shadow is about 100 to 150 km wide and the penumbra extends about 3000 km beyond the umbra. Locations within the Moon's umbra observe a total solar eclipse. Locations within the penumbra observe a partial solar eclipse. The eclipse path moves on the Earth's surface generally from west to east and sweeps a distance of 10 000 km over a period of 3–4 h. The eclipse starts from the west limb of the Sun's disk and proceeds towards the east. The actual totality lasts only a few minutes.

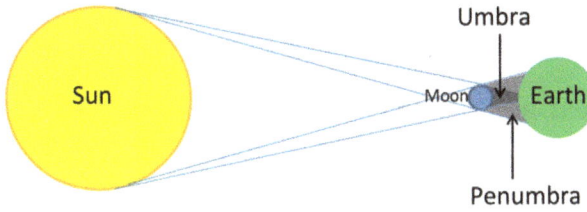

Figure 4.7. The geometry of a solar eclipse. The figure is not to scale.

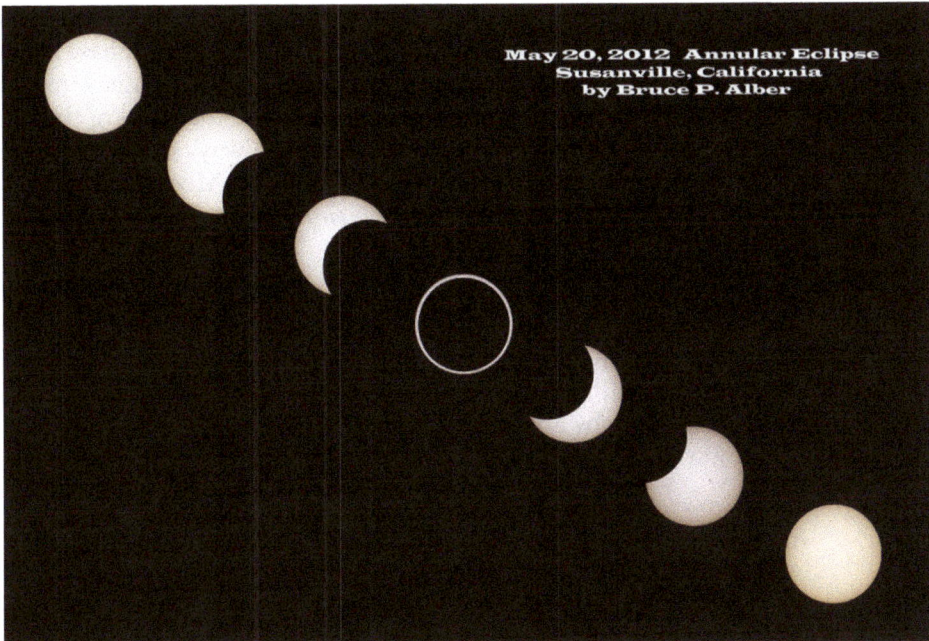

Figure 4.8. Annular solar eclipse of May 2012. The time sequence proceeds from left to right. Duration of sequence is about 3.5 hours. Copyright Bruce Alber. Used with permission.

Just before totality, the *chromosphere* (which is the lower atmosphere of the Sun) gradually becomes visible. At this point, the irregularities of lunar topography manifest themselves as bright spots in the ring of light surrounding the Moon's silhouette. The spots are the so-called *Baily's beads*. A single bright bead on the ring gives the *diamond ring* effect in a solar eclipse. Finally, during totality the observer has the unique opportunity to see the *corona*, which is the irregularly shaped outer atmosphere of the Sun, usually invisible in the glare of the Sun. If the solar eclipse happens when the Moon is not too close to the Earth, then the Moon's angular size[9] is not large enough to cover the entire Sun at totality, leaving a ring that is not covered by the Moon. The result is an *annular eclipse*. Figure 4.8 shows the

---

[9] See appendix A.

progression of an annular solar eclipse. A single eclipse may switch from annular to total. This is called a *hybrid eclipse*.

4.7 The sidereal and synodic periods of the Moon

Figure 4.9 shows the Moon's orbit around the Earth. At the top left position, the Moon and Sun are in the same direction and we have the new Moon phase. In the middle position, the Moon has completed one orbit around the Earth and, at the same time, one spin with respect to the stars (e.g. as seen by an observer in a distant spaceship). This is the *sidereal* (i.e. with respect to the stars) period. Meanwhile, the Earth has moved in its orbit around the Sun and it takes about two extra days for the Moon to line up in the direction of the Sun and arrive at the new Moon phase again. The orbital period and spin (the sidereal revolution and sidereal rotation, respectively) are both equal to 27.3 days. So from our vantage point, the effect of the spin cancels the effect of the orbital motion. The result is that we always see the same side of the Moon, as discussed in section 4.5. The interval from new Moon to new Moon is the *synodic* period (the synodic month or *lunar month*) and is 29.5 days long.

4.8 The apparent size of the Moon

The apparent size of the Moon, especially the full Moon, is not always the same. The size appears to change due mainly to the changing Earth–Moon distance. From the closest point in its orbit (the so-called *perigee*) to the farthest (the *apogee*), the Moon's apparent size changes by about 13%. In the popular media, a full Moon that occurs when the Moon is on or near perigee is referred to as *super Moon*. The change

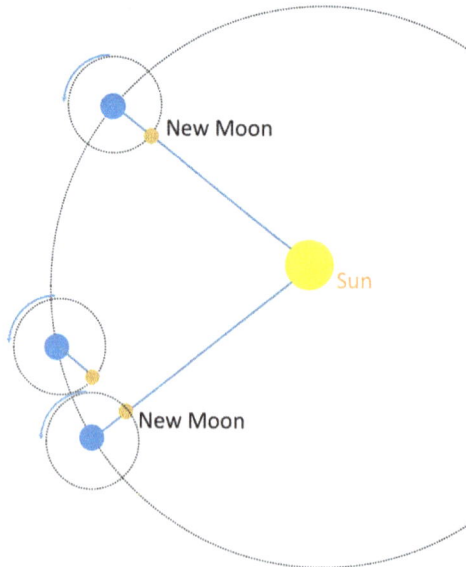

Figure 4.9. The synodic month of the Moon (from new Moon to new Moon). A complete rotation with respect to the stars is completed about 2 days earlier. This is the sidereal month of the Moon.

in size and brightness are not particularly striking, as the human visual system relies more on comparison (of size and of brightness).

A common experience is that the full Moon appears larger when closer to the horizon compared to when the Moon is high above the horizon. This is an illusion (the *Moon illusion*), and is an open topic of discussion. Using the practical angle measurement discussed in appendix A, one can verify that the Moon's apparent size is always 0.5°, more or less.

4.9 The ocean tides

The highs and lows in the ocean tides follow a close relationship with the Moon's phase. The moon's tidal effect is about twice as strong as the sun's effect. The exact timing and intensity depend strongly on the local topography, but as a general rule, the difference between successive highs and lows becomes largest during the new Moon and also during the full Moon phase. These are the *spring* tides. In the new Moon and full Moon phases, the Sun and the Moon line up with the Earth and their effects on the ocean tides are combined. During the quarter phases, the tidal effects of the Sun and Moon are at about 90° to each other, so they partially cancel each other. Consequently the difference between the successive high and low tides is small. These are the *neap* tides. During lunar and solar eclipses the alignment of the Earth–Sun–Moon is complete, which makes the spring tides during eclipses higher than normal. The effect of the planets on the ocean tides is small. Venus's effect is 19 000 times weaker than the tide caused by the Moon, and Jupiter's effect is about 160 000 smaller than the tide caused by the Moon. The effect of the rest of the planets is even weaker.

4.10 Resources

Information on past and future eclipses is available from http://eclipse.gsfc.nasa.gov/eclipse.html.

The phases of the Moon are available from http://asa.usno.navy.mil/.

Lunar calendars are available from a variety of sources online.

Chapter 5

The planets

The Greeks used the term *planets* (meaning 'wanderer' stars) to describe the irregular apparent motion of what we know today to be large objects orbiting the Sun. At the time, the known planets were the ones generally visible to the naked eye: Mercury, Venus, Mars, Jupiter and Saturn. Uranus was discovered much later (1781) and is at the limit of what the naked eye can see under ideal conditions. Neptune was discovered even later (1846) and is not visible to the naked eye. What is 'irregular' about the motion of the planets is not their diurnal (daily) motion in the sky. Just like the Sun and the Moon, the planets rise from the east and set in the west as observed from most locations on Earth (except the polar regions). On any given night, they appear to move in the sky at the same rate as the stars and their paths appear essentially parallel to the path of the stars. Also, just like the Sun and the Moon, the position of the planets relative to the stars changes with time.

With respect to the stars, the Moon shifts from west to east by about 13° every day. The Sun's daily shift relative to the stars is about 1°. For planets, the daily shift is about 2° for Mercury, 1° for Venus, less than 1° for Mars, and much less for Jupiter and the other planets. The shift is in the same direction (i.e. west to east) as the Sun and the Moon *most* of the time. The west to east shift relative to the stars is referred to as the *normal* or *prograde* apparent motion. What is irregular about the planets' motion is that occasionally the direction of the daily shift relative to the stars changes direction, i.e. they shift east to west. The east to west apparent path of the planets relative to the stars is referred to as *retrograde* motion. During retrograde motion, the planets rise and set as usual and the daily shift is comparable to the shift during prograde motion (i.e. 1° for Venus, less that 1° for Mars, etc). The duration of retrograde motion is different for different planets. For example, the retrograde motion of Venus (shown in figure 5.1(*a*)) lasts for about 6 weeks, and that of Mars (shown in figure 5.1(*b*)) lasts approximately 10 weeks. For Saturn the retrograde motion lasts about 20 weeks.

doi:10.1088/978-1-6270-5481-2ch5

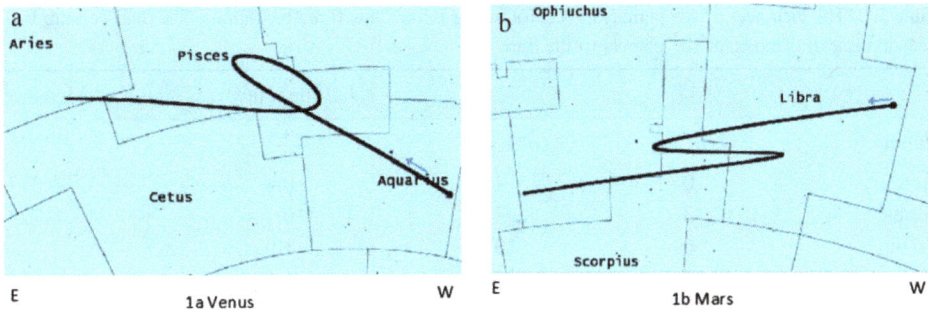

Figure 5.1. (*a*) The motion of Venus from 1 January 2017 to 15 June 2017. The retrograde motion (east to west) occurs from 5 March to 15 April 2017. (*b*) The motion of Mars from 1 January 2016 to 20 September 2016. The retrograde portion occurs from 20 April to 1 July 2016. The arrow indicates the direction of motion. Generated using Starry Night software.

The retrograde motion was easy to explain once it was realized that the Sun is the center of the Solar System (the heliocentric model) and the planets, including Earth, orbit around the Sun. We will return to the explanation of retrograde motion in section 5.6. First we will describe some orbital characteristics of planets.

5.1 The orbits of planets

All the planets orbit the Sun in the same direction. An observer looking at the Solar System from the north side (as we define north from Earth) would see all the planets orbit counterclockwise. The plane of the Earth's orbit is called the *ecliptic*. The orbits of all the planets are very near the ecliptic. Mercury's orbit has the largest inclination to the ecliptic, about 7°. As a result, with the exception of Mercury, all the planets (and the Moon) appear to orbit in a zone extending 5° above and below the ecliptic, called the *zodiac*.

The distance of planets from Earth is determined by radar. The method relies on reflecting a strong radio-signal. From the round-trip time of the 'echo' and knowing the speed of the radio signal (which is the speed of light) we can calculate the distance and from a series of distance measurements we can calculate the size of the planetary orbits. The final column in table 5.1 lists the distances of the planets from the Sun in astronomical units (AU). As the distances of the planets are not constant, the numbers listed in the table are the average between the maximum and minimum distance of the planet from the Sun. The average Earth–Sun distance is one AU, which is equal to 149 597 870.7 km and is used as a unit to express the distances of planets in the Solar System.

Bode's law (or the Titius–Bode law) is a rule that gives approximately the average distance of the planets to the Sun. The rule has no scientific basis and there is no evidence that it works in extrasolar planetary systems. The rule is based on the sequence of numbers 0, 3, 6, 12, 24, etc. After the first two numbers (0 and 3) to generate a number in the sequence we double the number preceding it. In table 5.1, the sequence is listed under the column heading *n*. Next we add 4 to each number in

Table 5.1. The distances of the planets in AU following Bode's law (fourth column). The final column is the measured mean distance of the planets to the Sun.

Planet	n	$n + 4$	Bode's law $\frac{n+4}{10}$	Measured a (in AU)
Mercury	0	4	0.4	0.317
Venus	3	7	0.7	0.723
Earth	6	10	1.0	1
Mars	12	16	1.6	1.524
(Ceres)	24	28	2.8	2.767
Jupiter	48	52	5.2	5.204
Saturn	96	100	10.0	9.582
Uranus	192	196	19.6	19.201
Neptune	384	388	38.8	30.047

the sequence and divide the sum by 10. The result, as listed in the 4th column of the table, is approximately equal to the measured mean distance of the planet from the Sun, a, listed in the final column of table 5.1 in AU.

It is interesting to note that the sequence included an empty spot between Mars and Jupiter, before the discovery of asteroids. The distance corresponds well to the orbit of the largest asteroid, Ceres (which was discovered 1801 and at present is considered a dwarf planet). It is also interesting to note that the sequence correctly predicted the distance of Uranus but was substantially off for the two discoveries that followed, i.e. Neptune and Pluto (not included in table 5.1).

From table 5.1 it is seen that the sizes of the orbits span a very wide range of values. The orbit of Neptune is about 80 times larger than the orbit of Mercury. If the orbit of Mercury is represented by a letter 'o' at the center of this page, then Neptune's orbit would extend into the margins of the page. One can then understand the difficulty of showing the planetary orbits in a diagram to scale[1]. Figures 5.2(a) and 5.2(b) show the planet orbits to scale.

5.2 Planetary motion and Kepler's laws

Since ancient times, and without any proof, the planetary orbits were assumed to be circular. Kepler was the first to challenge this assumption. Based on many years of observations, he concluded that the correct shape of the orbits is an ellipse. An ellipse has two foci and two axes, the major and the minor axis, as shown in figure 5.3. It is common to use half of the major and half of the minor axes (symbolized by a and b, respectively) to describe an ellipse. The difference between the two axes is a measure of how elongated the ellipse is. The *semi-major axis a* is the average between the maximum and the minimum distance from the ellipse to a focus of the ellipse. The following three laws describe the motion of the planets around the Sun.

---

[1] Most of the images displayed on the internet under 'Solar System to scale' etc, are actually not to scale.

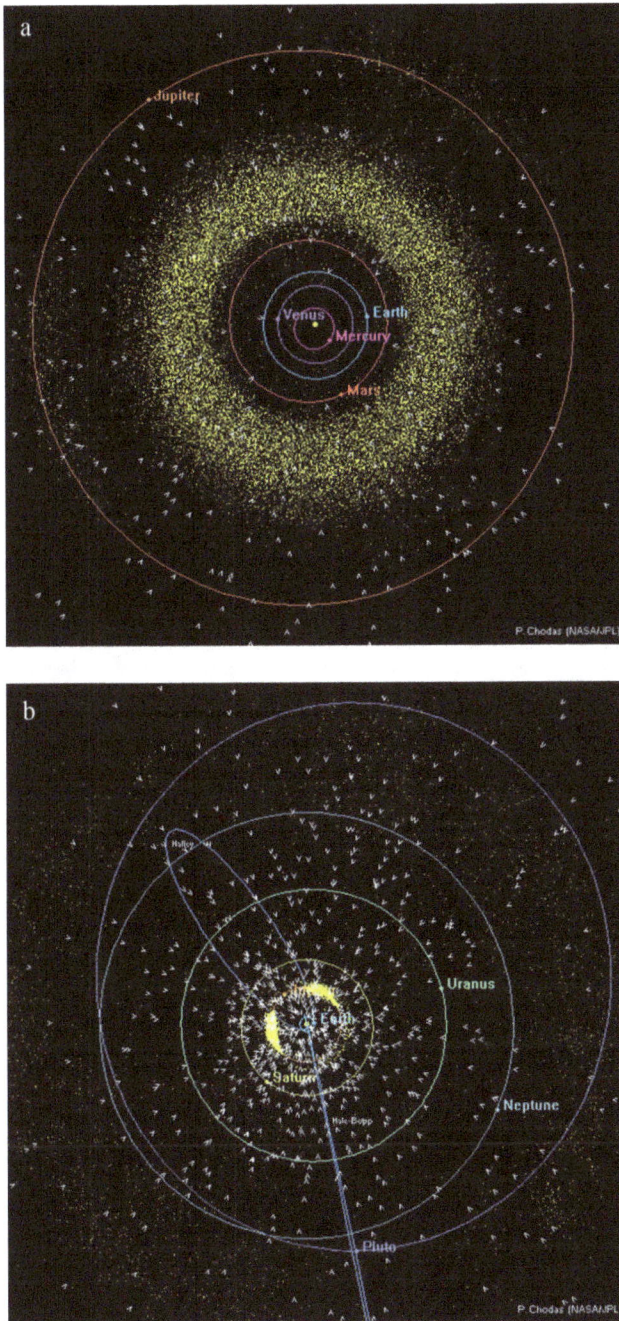

Figure 5.2. (*a*) Orbits of the first five planets (Mercury to Jupiter). The yellow dots indicate the main asteroid belt, between Mars and Jupiter. (*b*) Orbits of the outer planets and Pluto. The distances are to scale but sizes are not to scale. Credit: P Chodas NASA/JPL-Caltech, http://ssd.jpl.nasa.gov/?ss_inner and http://ssd.jpl.nasa. gov/?ss_outer, respectively.

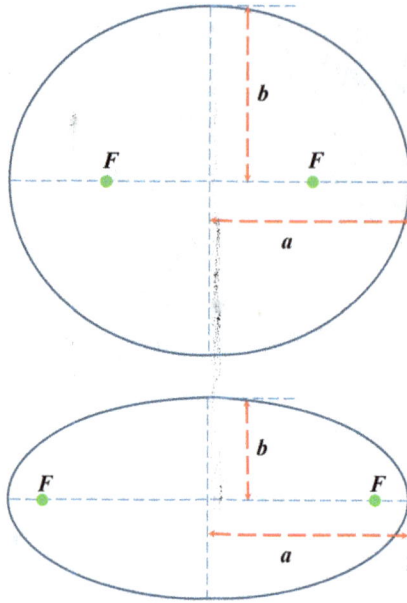

Figure 5.3. Two ellipses with equal semi-major axes, *a*. The bottom is more elongated (eccentric) because the difference between the semi-major axis, *a*, and the semi-minor axis, *b*, is larger. Note also that the two foci, marked by *F*, are further apart for the more eccentric ellipse.

5.2.1 Kepler's first law

The shape of a planet's orbit is an ellipse. The Sun is on one of the (two) foci of the ellipse. Because the orbit is an ellipse, the planet's distance from the Sun changes. The *perihelion* is the point when the planet is closest to the Sun. The *aphelion* is the point when the planet is farthest from the Sun. Both the perihelion and aphelion lie on the major axis of the ellipse. The semi-major axis, *a*, is equal to the average distance of the perihelion and the aphelion from the Sun. This is the distance listed in the final column of table 5.1 for each planet.

5.2.2 Kepler's second law

As the planets move in elliptical orbits, their distance to the Sun changes. The speed of the planet is not constant either. As the distance of the planet from the Sun changes, the speed of the planet changes. The speed of the planet changes in such a way as to keep the value of the product of the speed times the distance constant. As a result, the planet speeds up as it moves closer to the Sun and slows down as it moves farther from the Sun. A planet moves fastest around its perihelion and slowest around its aphelion. For Earth, perihelion is around 4 January and aphelion is around 4 July.

5.2.3 Kepler's third law

There is a relation between the time it takes a planet to complete one orbit around the Sun (i.e. the *period* of the orbit) and the size of the orbit. To measure the size of

Table 5.2. The orbital characteristics of the planets, http://nssdc.gsfc.nasa.gov/planetary/factsheet/.

Planet	Sidereal period P (Earth years)	Semi-major axis a (AU)	Average speed $(km\,s^{-1})^2$	Synodic period (Earth years)
Mercury	0.241	0.387	47	0.317
Venus	0.615	0.723	35	1.599
Earth	1	1	30	–
Mars	1.881	1.524	24	2.135
Jupiter	11.862	5.204	13	1.092
Saturn	29.457	9.582	10	1.035
Uranus	84.011	19.201	7	1.012
Neptune	164.79	30.047	5	1.006

the orbit we can use the semi-major axis, a. It is convenient to measure the period, P, in terms of Earth years and the semi-major axis, a, in AU. Kepler's third law states that for any planet:

$$a^3 = P^2.$$

Because of the exponents on each side of the above equation are different (cube and square) it follows mathematically that the planets move at different speeds and the outer planets move more slowly. The period refers to the *sidereal period*, i.e. the time it takes to complete an orbit with reference to the stars[3]. For example, for Mars $a = 1.524$ AU and $P = 1.881$ Earth years. We find $a^3 = 3.540$ and $P^2 = 3.538$, in good agreement with Kepler's prediction.

Table 5.2 lists some orbital characteristics of the planets. Note that the speed listed in the fourth column decreases the farther the planet is from the Sun, in agreement with Kepler's third law.

5.3 The phases of the planets

Just like the Moon, planets do not emit light of their own. They are visible only because they reflect sunlight. At any given time, one half of the planet is illuminated (the side that is facing the Sun) and the appearance of the planet from an Earth perspective depends on the direction we are viewing the planet. In other words, the planets show phases, just like the Moon. The phases of Venus were first observed by Galileo using a telescope. Some observers claim to be able to detect the crescent shape of Venus with the naked eye. Figure 5.4 shows the phases of Venus.

5.4 Planetary configurations

The location of the planets as seen from Earth can be described by the angle between the line of sight to the planet and the line of sight to the Sun. This angle is called the

---

[2] In standard scientific notation km s^{-1} means kilometers per second.
[3] The synodic period will be discussed in section 5.5.

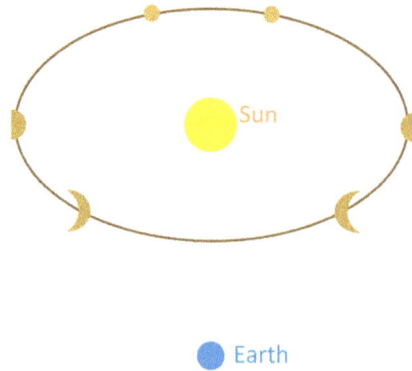

Figure 5.4. The phases of Venus. Note that during the crescent phases, Venus is very close to Earth and appears larger and brighter.

elongation of the planet, and was introduced in chapter 4. The elongation is measured in degrees, from 0° to 180° east of the Sun, and from 0° to 180° west of the Sun. Since the orbits of the planets are very near the ecliptic, we can measure the angles as projected on the ecliptic. Planets with orbits smaller than the Earth's orbit are classified as *interior* (also called *inferior*) planets (Mercury and Venus). Planets with orbits larger than the Earth's orbit are classified as *exterior* (also called *superior*) planets (Mars, Jupiter, Saturn, Uranus and Neptune).

When the elongation of a planet is 0° the planet is at *conjunction*, i.e. the planet approximately lines up with the Sun[4]. When an exterior planet is in conjunction, the Sun is between the planet and the Earth. For the interior planets there are two possibilities for conjunctions, the *superior* and the *inferior* conjunctions. The superior conjunction happens when the Sun is between the planet and the Earth. The inferior conjunction occurs when the planet is between the Sun and the Earth. In conjunction, the planets are in the same part of the sky as the Sun, therefore they are not visible because of the solar glare.

When the elongation of an exterior planet is 180° the planet is at *opposition*, i.e. the Sun and the planet are at opposite sides of the Earth. The interior planets *cannot be in opposition*. At opposition the planet is high in the sky at midnight and closer to the Earth. Therefore, opposition is *the most favorable time to observe an exterior planet*.

When the elongation of an exterior planet is 90° (east or west) the planet is at (east or west) *quadrature*. The *interior planets cannot be at quadrature*, since their elongation is always less than 90°. Specifically, for Mercury the *greatest* elongation is about 28° and for Venus about 48°. The *configurations* or *aspects* of exterior planets (Mars in this instance) are shown in figure 5.5.

The configurations of an interior planet (Venus in this instance) are shown in figure 5.6.

The best time to observe the interior planets is when their angular separation from the Sun is the largest. This occurs during the greatest elongations, shown in figure 5.7.

---

[4] The planet has the same ecliptic longitude as the Sun. See chapter 3.

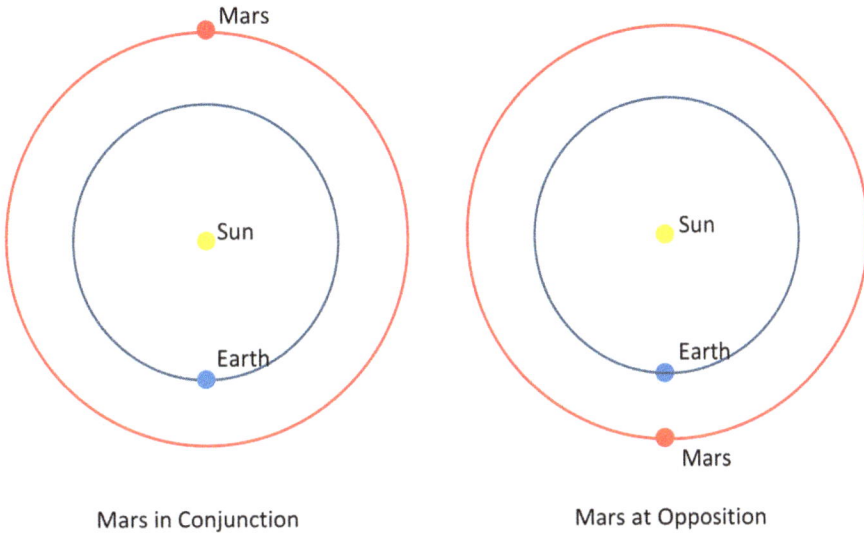

Figure 5.5. The configurations of an exterior planet (Mars).

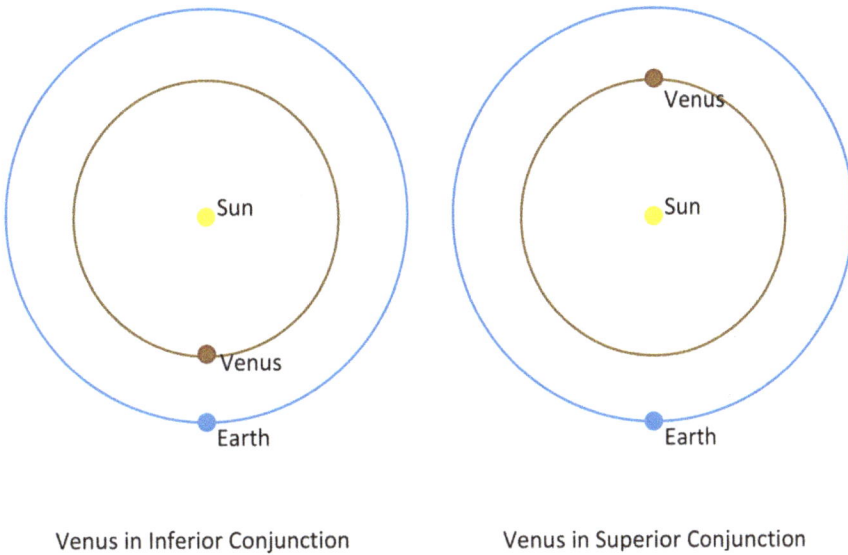

Figure 5.6. The configurations of an interior planet (Venus).

At greatest elongation the planet appears very bright. When Venus is at greatest elongation, the angular distance from the Sun is on average 48° (slightly over two hands at arm's length, as discussed in appendix A). Venus is at her brightest when the elongation is about 40° east or west. Note that as the Earth is spinning from west to east (counterclockwise in figure 5.7), when Venus is at western elongation she rises before the Sun and is visible at dawn in the east. Venus is then the *Morning Star*. When Venus is at eastern elongation, she is visible in the west after sunset and is the *Evening Star*.

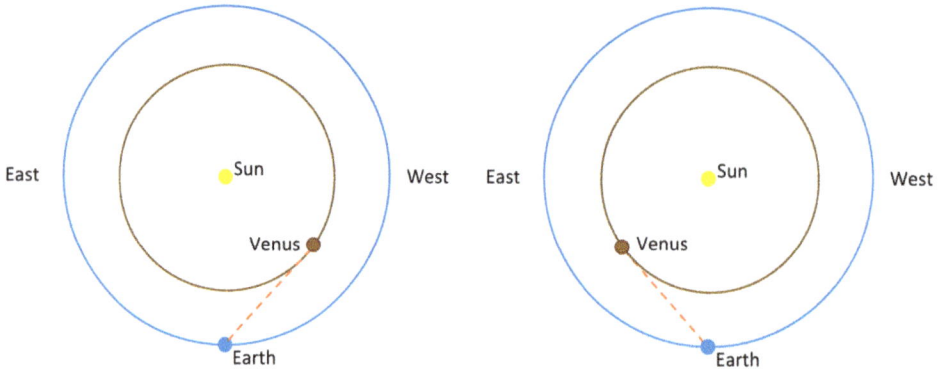

Figure 5.7. The greatest elongations of Venus. On the left is the greatest western elongation and on the right is the greatest eastern elongation. In this figure, the view is from north of the Solar System. The Earth is spinning counterclockwise when seen from the north.

5.5 Synodic and sidereal periods

The *synodic* period of a planet is the time it takes for a planet to come back to the same conjunction as seen from Earth. The synodic periods for planets are listed in the fifth column of table 5.2. For an interior planet, this would be the time it takes for the planet to gain one lap on the Earth. For an exterior planet it is the time it takes Earth to gain one lap on the planet. Looking at the synodic periods of table 5.2, we note that for the more distant planets (Jupiter, Saturn, etc) the synodic period is essentially equal to one Earth year. This happens because the Earth is moving much faster than the outer planets and gains one lap on them almost every year. This is not so for Mars or Venus, because their speeds are more comparable to the Earth's orbital speed, therefore it takes longer to gain one lap.

Tables 5.3, 5.4 and 5.5 list the dates of a few successive configurations for Mercury, Venus and Mars. The tables also list the elapsed time between successive configurations in Earth days. A quick glance at the number of elapsed days between successive configurations shows that there is a fluctuation of a few days, which means that the synodic periods of the planets are not constant, but change by a few days with every cycle. The synodic periods listed in the final column of table 5.2 are average values. The synodic period of planets fluctuates because the orbits of the planets are not circular and the speeds of the planets are not constant, as predicted by Kepler's laws.

Knowing the synodic period allows us to determine the dates favorable for observing a planet. For example, if Mars is today at opposition (favorable for observation) it will be in good position for observation again in 2.1 years (about 25 months). Or, if Mars is in conjunction (a bad time for observation) it will be in opposition in half its synodic period (about 13 months).

The *sidereal period* of a planet is the time it takes to complete an orbit around the Sun with reference to the stars. To measure the sidereal period, the observer should be stationary with respect to the stars. An observer on Earth is orbiting the Sun and

Table 5.3. The conjunction dates and elapsed time for Mercury.

Conjunction	Date	Days elapsed
Inferior	30/05/2015	NA
Superior	23/07/2015	54
Inferior	30/09/2015	69
Superior	17/11/2015	48
Inferior	14/01/2016	58

Source: http://ntserver.ct.astro.it/cgiplan/.

Table 5.4. The conjunction dates and elapsed time for Venus.

Conjunction	Date	Days elapsed
Inferior	15/08/2015	NA
Superior	06/06/2016	296
Inferior	25/03/2017	292
Superior	09/01/2018	290
Inferior	26/10/2018	290

Source: http://ntserver.ct.astro.it/cgiplan/.

Table 5.5. The conjunction and opposition dates, and elapsed time for Mars.

Configuration	Date	Days elapsed
Conjunction	14/06/2015	NA
Opposition	22/05/2016	343
Conjunction	26/07/2017	430
Opposition	27/07/2018	366
Conjunction	02/09/2019	402

Source: http://ntserver.ct.astro.it/cgiplan/.

therefore is not stationary. What an Earth-based observer can measure is the *synodic period*. The relation between the sidereal period (P_{si}) and synodic period (P_{sy}) is different for interior and exterior planets. For the calculations we express the periods in terms of Earth years.

For *interior* planets the relation between the sidereal period and synodic period is:

$$P_{si} = P_{sy}/\left(P_{sy} + 1\right).$$

For example, for Venus $P_{sy} = 1.60$ years. Therefore,

$$P_{si} = P_{sy}/\left(P_{sy} + 1\right) = 1.60/(1.60 + 1) = 1.6/2.6 = 0.615 \text{ years.}$$

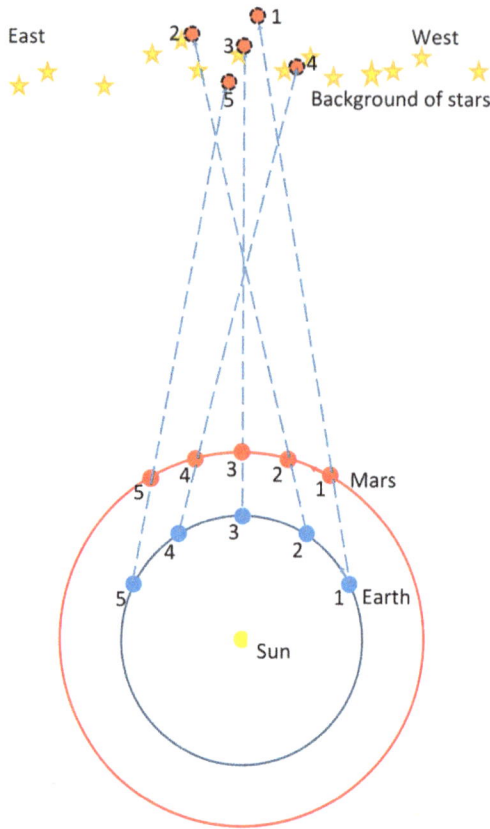

Figure 5.8. An explanation of the retrograde motion of an exterior planet, in this instance Mars. Earth is moving faster than Mars. When Earth is overtaking Mars (positions 2–4) the apparent position of Mars in the background of stars shifts from east to west. This is the retrograde part of the apparent motion of Mars. Drawing not to scale.

For *exterior* planets the relation between the sidereal period and synodic period is:

$$P_{si} = P_{sy}/(P_{sy} - 1).$$

For example, for Mars $P_{sy} = 2.14$ years. Therefore,

$$P_{si} = P_{sy}/(P_{sy} - 1) = 2.14/(2.14 - 1) = 2.14/1.14 = 1.88 \text{ years}.$$

5.6 An explanation of the retrograde motion of planets

Using the relation between distance and period, the retrograde motion of the planets can be explained quite simply. Consider for example the case of Mars. As Mars is an exterior planet, it moves more slowly on its orbit compared to Earth. Remember that retrograde motion is from east to west *with respect to the stars*. Figure 5.8 shows how Mars appears in relation to the stars, as the Earth is overtaking Mars on its orbit.

The figure shows different positions taken at set time intervals (approximately every month).

As the Earth shifts from position 1 through to position 5, so does Mars, the only difference being that the Earth moves faster and travels a larger distance in each step. The relative position of Mars in the background of the stars is shown by the extension of our line of sight (the dashed lines). From positions 1 to 2, the line of sight shifts from west to east relative to the background of stars. This is called direct or prograde motion. From positions 2 through 4 the line of sight shifts from east to west. This is called retrograde motion. From positions 4 to 5 the line of sight shifts from west to east again and prograde motion resumes. The retrograde motion happens when the Earth overtakes Mars (positions 2, 3 and 4) and this is when Mars is at opposition (see figure 5.5). A similar diagram could be constructed to explain the retrograde motion of interior planets. For the interior planets the retrograde motion occurs when the interior planet is overtaking Earth on its orbit and this corresponds to the time of an inferior conjunction.

5.7 Observing the planets

Locating the planets in the night sky with the naked eye and observing the planets through binoculars are some of the most rewarding experiences for sky watchers.

Favorable conditions for observing *Mercury* only occur a couple of times per year around sunrise or sunset, because the planet is always so close to the Sun. In addition, the inclination of the orbit relative to the horizon of the observer (see figures 3.5–3.7) may keep the planet in the glare of the Sun, even when the planet is at greatest elongations. Finally, Mercury lacks any distinctive surface features such as, for instance, the Moon's maria (see chapter 4), which makes it difficult to explore the topography.

Venus lacks visible surface features because the planet is surrounded by thick cloud cover. Because the clouds reflect the sunlight very effectively, Venus appears very bright. Venus can be visible to the naked eye even in the daytime. A pair of binoculars or a small telescope will reveal the phases of Venus and more impressively the crescent phases shown in figure 5.4.

Mars is best observed in opposition, which is the most favorable configuration for observing the exterior planets. Mars has a characteristic red color and is easy to spot. With a small telescope under good conditions, one can distinguish features of the Martian surface, including the *polar ice caps*.

Jupiter has four large satellites, the so-called *Galilean moons*, first observed by Galileo. Some observers claim to be able to see the Galilean moons with the naked eye. The Galilean moons are certainly clearly observable with binoculars or a small telescope. A small telescope will also show the color band appearance of Jupiter's thick atmosphere, as well as the *Giant Red Spot* just south of Jupiter's equator.

Probably the most fascinating view of all is the view of *Saturn's rings*. All the large planets (Jupiter, Saturn, Uranus and Neptune) have rings, but Saturn's are the most extensive and are visible from Earth even with a small telescope. Observing the different rings and the gaps between them requires more than a small telescope yet, even at the smallest resolution, the scene is magnificent. The rings of Saturn are inclined to the orbit and their visibility from Earth goes through cycles of about 15 years. The appearance of the rings should be at its best around 2017–18.

Observing *Uranus* is more of a challenge. The planet's brightness is at the limit of what can be seen with the naked eye. Through a telescope, the planet appears bluish and lacks any distinctive features. Uranus is covered by a thick atmosphere but does not show any color bands like Jupiter.

5.8 Resources

There are several resources for locating planets.

Astronomy magazines post online weekly information on the location of planets, stars and other useful information. For example, *Sky and Telescope* (skyandtelescope.com) posts 'This Week's Sky at a Glance'; the *Astronomy Magazine* (astronomy.com) posts 'The Sky This Week'.

Sky simulation software: some are available as free downloads (Stellarium) and some can be purchased for a range of prices (MyStars, Starry Night, etc).

Astronomical almanacs, such as asa.usno.navy.mil.

Coordinates can be downloaded as an 'ephemeris' for example, from http://ssd.jpl.nasa.gov/horizons.cgi - top.

Visual Astronomy
A guide to understanding the night sky
Panos Photinos

Chapter 6

Comets, meteoroids and meteor showers

The larger objects in the Solar System orbit on or near the ecliptic and in the same direction as the Earth[1]. Comets and other smaller objects are often the exception to this rule. The orbits of comets can have any direction and any inclination to the ecliptic. The word 'comet' has a Greek origin which means 'hairy', describing the 'fuzzy' appearance of these objects. The Greeks described them as 'hairy stars'.

6.1 The structure of comets

Comets are icy objects with a typical size of a few tens of kilometers. At their brightest, comets develop tails that can be tens of millions of kilometers long, making them the most extended objects in the Solar System. The material forming the comet's icy nucleus is leftover material from the formation of the Solar System that was tossed to the outer parts of the Solar System where the low temperatures allowed them to survive for billions of years. The ice is primarily water ice, with smaller proportions of ammonia, carbon dioxide and carbon monoxide ices. The nucleus has a dark crust and a porous internal structure. The composition of the nucleus includes silicon, some metals and some organic material, in proportions which are consistent with the composition of the early Solar System.

Occasionally, a gravitational interaction with another object disturbs the comet's orbit enough to change its course towards the inner Solar System. There the solar heat causes the ices inside the nucleus to vaporize. The vapors (H_2O, CO_2, etc) leak through the crust as *jets*. As they exit, the jets push along tiny dust particles. The jet material creates a gas and dust envelope around the nucleus called the *coma*. When first spotted, the comets have a fuzzy appearance, which is due to the coma. The size of the coma is typically of the order of 100 000 km, but can vary significantly with time. The coma and the icy nucleus form the *head* of the comet. A hydrogen

---

[1] See chapter 5.

6-1

envelope surrounds the coma, extending to several million kilometers. This envelope emits ultraviolet radiation, which is not visible to the unaided eye.

When the comet comes within approximately 2 astronomical units (AU) from the Sun, the coma experiences the effects of light and particles (protons and electrons that form the so-called *solar wind*) coming from the Sun. The radiation is so intense that it creates a pressure pushing outwards (i.e. away from the Sun). The charged particles in the solar wind also contribute to the outward push. These particles are moving outwards at speeds exceeding $500 \, km \, s^{-1}$. As a result, the dust and gas is pushed away from the coma, forming the dust and gas tails that point away from the Sun. The *gas tail* consists of atoms and molecules, which are small and are aligned quickly in the direction of the outward push, before the head has had the time to move significantly along its orbit. Thus the gas tail points directly outward from the head. The *dust tail* consists of smoke sized particles, each containing trillions of molecules. Having a larger mass, the dust particles move outward much more slowly. In the meantime the head has moved ahead along its orbit and the dust is lagging behind. As the push diminishes outward, the result is a curved dust tail. The dust tail is visible because the dust particles reflect sunlight. It is the most distinct feature visible to the naked eye and in telescope images it appears yellowish in color. In addition to being curved, the dust tail is often highly textured, as shown in figure 6.1.

The *gas* or *ion tail* is straight, and consists mainly of water, carbon monoxide (CO) and cyanogen (CN). Under the influence of the intense solar radiation the neutral molecules in the gas may become ionized, i.e. electrically charged (hence the ion tail). The ions can emit light by recombining with electrical charges to become neutral again. The light emitted in this process by the carbon monoxide (CO^-) ion

Figure 6.1. The tail of Comet McNaught over the Pacific. January 2007. Credit: ESO/H H Heyer, http://www.eso.org/public/images/img_4981/.

usually gives the bluish color to the ion tail as seen in telescope images. The electrical charges of the ion tail interact with the electrical charges in the solar wind, giving the ion tail its straight shape. Telescopic observations of the ion tail reveal kinks, disruptions and, at times, separation of the tail. These features reflect the dynamic interaction of the charges in the ion tail with the solar wind. The density of the material around the icy nucleus is much lower than the Earth's atmosphere. Therefore, it is not surprising that stars can be visible through the head and tails of the comet, as shown in figure 6.2.

The appearance of the comet depends on the angle of view and the location of the observer with respect to the plane of the comet's orbit. It is important to keep in mind that the main feature observed by the unaided eye is the dust tail and that the dust is spread like a fan on the plane of the comet's orbit. If the view is broad side, we see light coming from a wider but thinner layer of dust. If we have an edge-on view, we see a narrower but thicker layer of bright dust, so the tail will appear brighter and longer. This situation occurs when the Earth crosses the plane of the comet's orbit and, under certain conditions, the tail may appear to extend on both sides of the head. The result is an *anti-tail* which appears to point from the head towards the Sun. Figure 6.3 shows the anti-tail of Comet Hale–Bopp.

The direction of view may affect the appearance of the comet in other ways as well. For example, if the direction of the comet's motion is close to our line of sight, and the comet is approaching us, then the tails may appear shorter. If the comet is

Figure 6.2. Comet West was discovered in photographs by Richard West on 10 August 1975. It reached peak brightness in March 1976. During its peak brightness, observers reported that it was bright enough to study during full daylight. The comet has an estimated orbital period of 558 000 years. Note that stars are visible through the dust tail. Credit: J Linder/ESO, http://www.eso.org/public/images/c-west-1976-ps/.

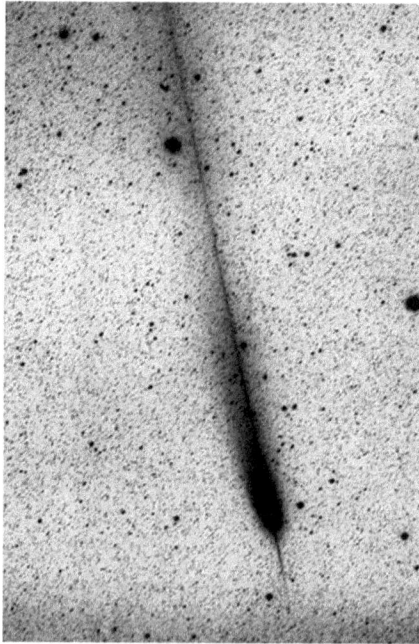

Figure 6.3. Comet Hale–Bopp showing an anti-tail. Photographed on 5 January 1998. Credit: ESO, http://www.eso.org/public/images/eso9806b/.

moving away from us (the tail is closer to the Earth than the head) the tail may appear much wider, like a fan. This is a familiar effect of perspective that makes the gap of railroad tracks appear narrower with distance (see section 6.4).

The brightness of the comet is determined by the size of the nucleus, the shape of the orbit and the location of the Earth during the weeks/months preceding and following the comet's nearest approach to the Sun (the so-called *perihelion*). The nucleus of Comet Hale–Bopp was 60 km on average (compared to 11 km for Halley's Comet), the orbit was inclined and the perihelion occurred well above the ecliptic. All these facts combined to make the comet visible for 18 months. Presumably the tails would be the largest at perihelion. However, if the perihelion is along our line of sight to the Sun, the proximity of the comet to the Sun makes observation unfavorable, because the comet would be hidden by the Sun's glare. In the last visit (the so-called *apparition*) of Halley's Comet (1986) the conditions were particularly unfavorable, since at perihelion, the Sun was between the Earth and the comet, and the comet was visible at intervals of a few days at a time from January to April of 1986.

6.2 The orbits of comets

Like all objects in the Solar System, comet orbits follow Kepler's laws, discussed in chapter 5. By observing the comet's path at each apparition, astronomers can determine the shape of the entire orbit of the comet. The orbits of comets are highly *eccentric*, meaning that their closest approach (the *perihelion*) is much smaller than their largest distance from the Sun (the *aphelion*). For example, for Halley's Comet, the perihelion is only 0.59 AU from the Sun while the aphelion is 35.2 AU. For other comets the aphelion can be at tens of thousands of AU from the Sun.

The semi-major axis *a* of the comet's orbit is equal to the average distance of the perihelion and the aphelion from the Sun. For Halley's Comet $a = (0.59 + 35.2)/2 = 17.9$ AU. Knowing *a*, we can calculate the period *P* of the comet from Kepler's third law, which states that[2]

$$a^3 = P^2.$$

For Halley's Comet we have $P^2 = 17.9^3$ and using a calculator we find the period $P = 76$ years. Using this method we can calculate the period of the comet's orbit, even if the comet is seen only once.

As a result of the high eccentricity of comet orbits, the speed of the comet near the perihelion is much higher than the speed at aphelion. This follows from Kepler's second law[3]. For example, near perihelion, Halley's Comet is moving at approximately 55 km s^{-1} (kilometers per second). At aphelion the speed drops to less than 1 km s^{-1}. Because of the higher speed, the comet's travel time through the inner Solar System is a small fraction of the orbital period. For instance, Halley's comet is closer than 1.5 AU from the Sun (within the orbit of Mars) for approximately 5 ½ weeks out of its 76 year period.

6.3 Long and short period comets

Depending on their orbital period, comets are classified as *short* or *long period comets*. Short period comets have periods less than 200 years. For example, Halley's Comet is a short period comet, because it has a period of 76 years. The short period comets, with many other icy bodies, orbit in the *Kuiper belt*, which is a region on the ecliptic extending from about 30 to 55 AU from the Sun. As a comet passes through its perihelion, approximately 1% of the mass of the nucleus is lost in the form of gas and dust tails. Eventually, the loss from repeated apparitions may not leave enough ices in the nucleus to generate jets or tails and the comet becomes extinct (dead). While the comet may no longer be visible, the dust lost by the comet continues to orbit the Sun and can reappear in the form of *meteor showers*, as discussed in section 6.4.

At the rate of a 1% loss with every apparition, all the short period comets become extinct after a few hundred apparitions. Keeping in mind that comets formed 4.5 billion years ago (the estimated age of the Solar System), the fact that short period comets still exist after 4.5 billion years indicates that there is a reservoir of comets somewhere in the Solar System. The *Oort cloud* is hypothesized to be the home of probably trillions of comets. The cloud extends from about 5000 to 100 000 AU and is more or less spherical. The spherical shape is consistent with the fact that long period comets can come from any direction. A comet from the Oort cloud would be a long period comet. If the comet wanders into the inner part of the Solar System, its

---

[2] See chapter 5.
[3] See chapter 5.

orbit may be altered by gravitational interaction with the larger planets, most probably Jupiter, and become a short period comet.

In addition to the loss of mass at each apparition, comets may fragment under the gravitational stretch they experience from the Sun or from a planet. Comet West came closer than 0.2 AU at perihelion (1976) and was observed to fragment shortly after perihelion. Comet Shoemaker–Levy 9 was captured into orbit by Jupiter. Eventually the comet fragmented and the fragments plunged into Jupiter in 1994 (http://www.eso.org/public/images/eso9410a/). Comet ISON had a perihelion of just over 0.01 AU and apparently fragmented upon reaching perihelion and the fragments disappeared in the Sun in 2013 (http://sohowww.nascom.nasa.gov/hotshots/index.html/).

6.4 Meteors and meteor showers

Interplanetary space is littered with *debris* of various sizes. The debris orbit in all directions and in all inclinations to the ecliptic. There are trillions of such objects of various sizes orbiting the Sun. Some of these objects are fragments from collisions between asteroids. Others fragmented from planets or the Moon during impacts with large objects. Such collisions were frequent in the early stages of the formation of planets. Another contributor is the dust that a comet leaves in its orbit, which becomes significant when the comet crosses its perihelion.

The term *meteoroid* refers to debris that wander close enough to the Earth and are eventually pulled down by the Earth's gravity. During this process, meteoroids acquire very high speeds, typically more than $10 \, \text{km s}^{-1}$ and often over $50 \, \text{km s}^{-1}$ ($180\,000 \, \text{km h}^{-1}$). As the object descends in the atmosphere, it slows down by friction with the atmospheric gas. The heat generated by friction vaporizes the descending object and heats up the surrounding gas. The result is a *meteor*, i.e. a brief light streak across the night sky. Larger objects can create a streak bright enough to be seen in the daytime and are called *fireballs*.

The majority of meteors are caused by tiny particles falling through the Earth's atmosphere. The bright streaks occur at a height between 50 and 130 km above the Earth's surface, on average. The height can be determined by observing the meteor from two different locations separated by 20 km or so. The light streak lasts from a fraction of a second to about 5 s, depending of the size of the particle. The total path on average is about 300 km long. If the particles are small, they burn out completely in the process.

The term '*shooting star*' is used as a synonym for meteor, although stars are not involved in this phenomenon. If the falling object is large enough, it may survive the high temperature trip through the atmosphere and land as a rock on the Earth's surface. In this case we speak of a *meteorite*. It is estimated that the total material that falls on Earth is about 1000 tons per day. Most of this material is in the form of dust. Occasionally, large meteorites are tracked falling (*falls*), or simply found on the ground (*finds*). Some meteorites (the *irons*) contain significant amounts of iron and other metals which makes them easy to distinguish from surrounding ordinary rocks. Others appear very similar to Earth rock (the *stones*) and are harder to spot.

In Antarctica all three types of meteorites, irons, stones and *stony irons*, which is the intermediate type, are readily distinguished in the icy surroundings.

On Earth, there are several impact craters that mark the sites of meteorite impacts and undoubtedly many more impacts occurred but their craters eroded away with time. The Vredefort crater in South Africa is the largest known impact crater on Earth. The crater is old (over 2 billion years) and eroded. Its original diameter is estimated at about 300 km and was caused by an object of about 10 km (http://www.vredefortdome.org). There is no visible meteorite on the crater site. It is interesting to point out that the Hoba meteorite in Namibia, which at 66 tons is the largest single piece known in the world, is not surrounded by an impact crater!

One of the most spectacular events recorded recently is the meteorite that fell in the Chelyabinsk region of Russia, near the Kazakhstan border. The meteorite (0.6 m in size, weighing 650 kg) was eventually recovered from the bottom of a nearby lake. The original object must have been around 20 meters and weighed over 10 000 tons. The meteor (a *super fireball*, as it was classified) appeared brighter than the Sun (http://www.nasa.gov/content/goddard/around-the-world-in-4-days-nasa-tracks-chelyabinsk-meteor-plume/#.VITOmIfahl0).

At certain times of the year, meteors appear from a certain point in the sky. These are the *meteor showers* and they appear to radiate from one point in the sky (the *radiant* of the meteor shower). The paths of the meteors are coming from one small region and are more or less parallel to each other. They appear to radiate from a point because of perspective, much like the railroad tracks shown in figure 6.4.

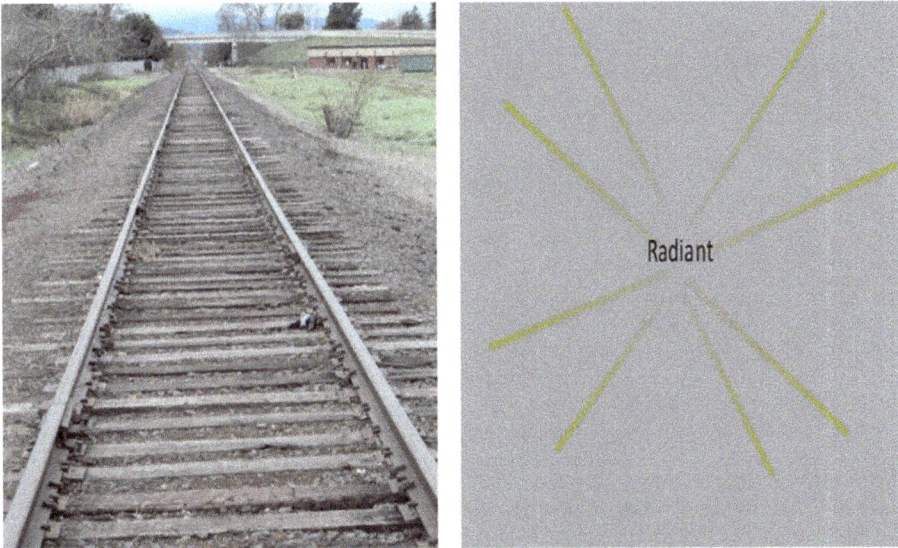

Figure 6.4. The effects of perspective. The gap between the rails appears narrower with distance (left). In the same way, the separation between meteors (right) appears to diminish with distance and they appear to originate from one point, the radiant of the shower.

Meteor showers occur when the Earth crosses the path of a comet. The dust left by the comet in its path causes these meteors. The different showers repeat at the same time of the year and are named for the constellation where the radiant of the meteor shower is located. For example, the radiant of the meteor shower occurring in the latter part of October is in the constellation of Orion, hence the Orionids. In some cases the comet can be determined, for example, the Orionids occur when the Earth crosses the path of Halley's Comet. In some cases the comet is unknown, for example the Quandrantids, occurring in early January.

The intensity of a meteor shower can vary from year to year. The shower can be very unimpressive for several years and then can be very spectacular, with many thousands of meteors per hour. For example, the Leonids appear to have a maximum every 33 years. The parent comet for this shower is Comet Tempel–Tuttle, and the maximum follows the passage of the comet through the inner Solar System on its 33 year orbital period.

As the Earth's average speed is over $107\,000$ km h^{-1},[4] the Earth can cross the dusty trail of a comet in less than 24 h. Therefore in some locations the height of the peak of the meteor shower is not visible because it occurs during daytime. The exact time of entry cannot be predicted precisely. What is certain is that the particles are more likely to hit the side of the Earth that happens to be in the forward direction of the Earth's motion at the time of the encounter with the swarm of particles in the dusty comet trail. As indicated in figure 6.5, the Sun is about to rise above the

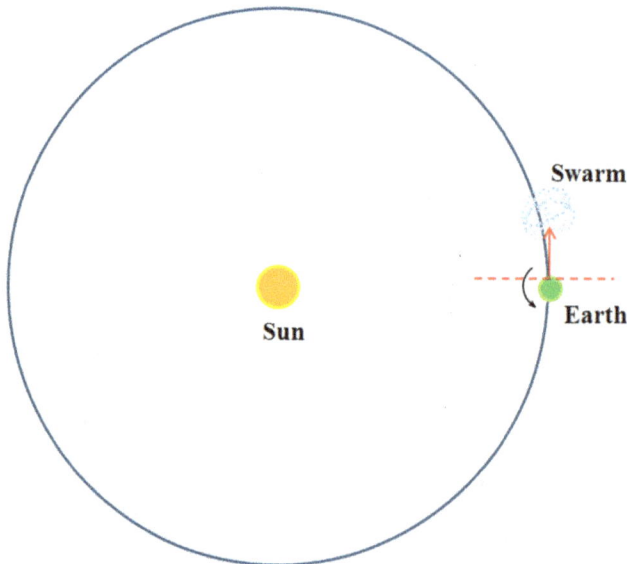

Figure 6.5. Earth crossing a swarm of particles. The forward direction is indicated by the red arrow. The dotted line is the horizon at the locations that happen to be in the forward direction. The Sun is about to rise at these locations. The curved arrow indicates the direction of the Earth's rotation.

---

[4] In standard scientific notation, km h^{-1} means kilometers per hour.

Table 6.1. The major meteor showers.

Name of meteor shower	Peak dates
Quadrantids	3/4 January
Lyrids	21/22 April
Perseids	12/13 August
Orionids	21/22 October
Leonids	17/18 November
Geminids	13/14 December

horizon at the locations that happen to be on the forward side. Table 6.1 shows a list of the major meteor showers.

6.5 Space junk

The space exploration missions and the deployment of satellites created a new type of debris orbiting the Earth. NASA estimates that approximately 19 000 man-made objects larger than 10 cm orbit the Earth as of July 2009. These include parts from spacecraft launches, old communication satellites, etc. They are referred to as *space junk* and at times a bright path resembling a fireball marks their descent, (http://earthobservatory.nasa.gov/IOTD/view.php?id=40173).

6.6 Resources

The properties of comets can be found at http://nssdc.gsfc.nasa.gov/planetary/planets/cometpage.html.

Visual Astronomy
A guide to understanding the night sky
Panos Photinos

Chapter 7

Constellations, asterisms and star names

The origin of the constellations dates back to at least 3000 BC, and it appears that they were part of the process of marking directions, time and seasons. The star patterns are outlined by 'stick-figures' connecting the brightest stars in the group. The figures are seldom suggestive of the name of the constellation. Nevertheless, the constellations provide a practical way to identify portions of the sky, to locate stars and to name stars in the so-called Bayer system, which is discussed in section 7.3.

Today the International Astronomical Union (IAU) recognizes 88 astronomical constellations of different sizes. The list includes many of the ancient patterns and those of the Southern Hemisphere introduced in the 16th century and onwards. The IAU definition provides boundaries which do not follow the familiar figures associated with the constellation. The boundaries are drawn nearly parallel to the declination and right ascension (RA) lines. The boundaries can be rather complex at times. Figure 7.1 shows part of the constellation of Serpens (the snake). Serpens is a constellation that consists of two parts, Serpens Caput (the snake's head) and Serpens Cauda (the snake's tail). They are located on either side of Ophiuchus (the serpent bearer). The RA and declination lines are marked by hours and degrees, respectively. The boundaries follow the directions of the RA and declination lines, but the overall pattern is complex.

Asterisms are patterns formed by groups of stars within a single constellation, or patterns formed by stars belonging to different constellations. Examples of asterisms formed by stars within a constellation are the Northern Cross (in the constellation of Cygnus), the Southern Cross (in the constellation of Crux) and Orion's Belt. Examples of asterisms formed by stars in different constellations are the Summer Triangle (Deneb in Cygnus, Altair in Aquila and Vega in Lyra) and the Winter Triangle (Sirius in Canis Major, Procyon in Canis Minor and Betelgeuse in Orion). Asterisms can vary widely in size. For example, the Pleiades (the Seven Sisters) in Taurus span a little less than 1°, the Big Dipper's long dimension is about 26° and the Summer Triangle's height is about 38°.

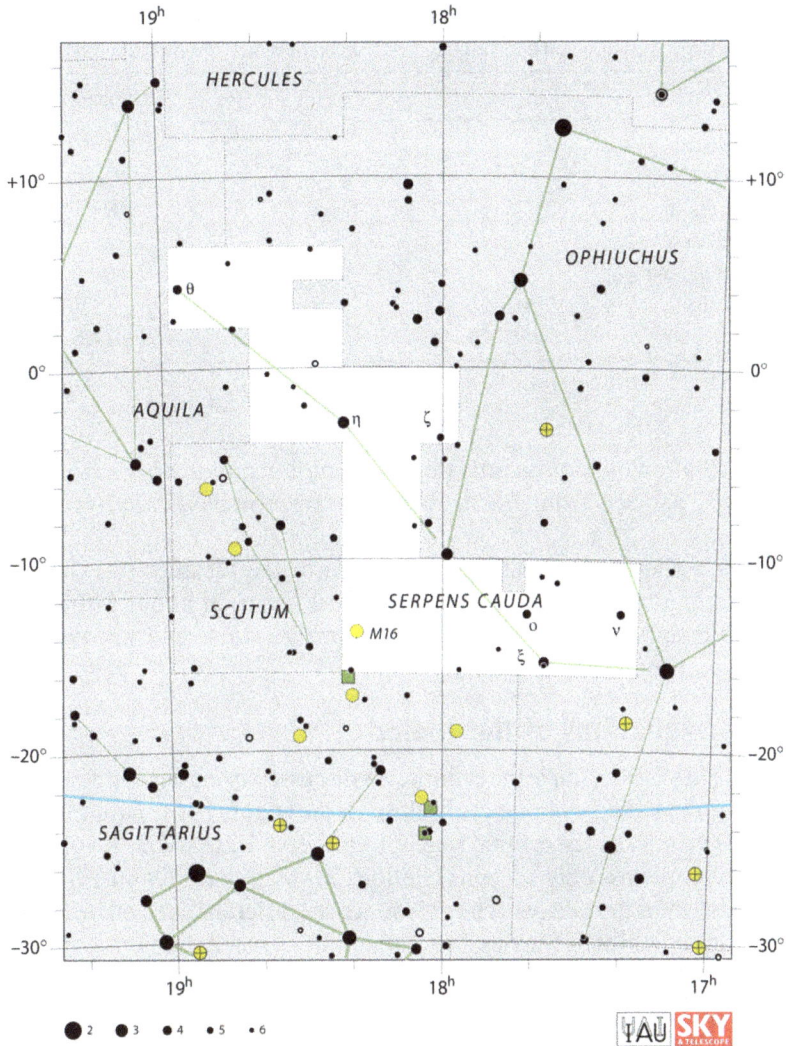

Figure 7.1. The east part of the constellation of Serpens (the snake). Note that the boundary lines follow the RA and declination lines. Credit: International Astronomical Union, and *Sky and Telescope*.

7.1 The distances between stars in constellations and asterisms

The stars forming the constellation and asterism patterns look 'close' together like a group only from our Earth-based perspective. Actually, the stars that make a constellation may be at enormous distances from each other. Table 7.1 lists the distances of several stars from the Sun in light years (ly). The first three stars in the table are the stars of Orion's Belt. The fourth star (Bellatrix) is also in the constellation of Orion. Note that Bellatrix is much closer to the Sun than it is to the stars in Orion's Belt. The last three stars in the table are the stars of the Winter Triangle and again we note considerable difference in the distances. Therefore, as a general rule, the

Table 7.1. A list of selected stars and their distance from the Sun. The distances are derived from parallax data using http://simbad.u-strasbg.fr/simbad/sim-fid.

Star name	Distance (ly)
Alnilam	1976
Alnitak	736
Mintaka	692
Bellatrix	252
Betelgeuse	498
Sirius	8.6
Procyon	11.5

stars in constellations and asterisms do not form a group in space. They are in the same direction, as seen from Earth, but they are not physically (gravitationally) connected.

A major exception to the rule is the asterism of the Pleiades (the Seven Sisters), which is a group of stars (a few hundred, not just seven) at about 440 ly away from the Sun. These stars are part of an *open cluster* of stars, and apparently formed approximately at the same time as a group.

7.2 The constellations of the zodiac

The apparent path of the Sun is the ecliptic. As discussed in earlier chapters, the path of the Moon and the naked eye planets is close to the ecliptic. It is natural that the star patterns in the vicinity of the ecliptic were the earliest to be marked. It appears that the most ancient records list only six constellations in the zodiac: Taurus, Cancer, Virgo, Scorpio, Capricornus and Pisces. The reason we see different constellations during the year is that the Earth orbits around the Sun. The constellations that are along our line of sight to the Sun are in the sky during the day so they are not visible. The constellations that are in the opposite direction are visible at night. For example, during the month of June, the Sun is along our line of sight to the constellation of Taurus, therefore Taurus is up during daytime in June, while Scorpius, which is in the opposite direction from the Sun from our perspective, is visible in the night sky in June.

The dates listed in horoscopes do not actually correspond to the time when the Sun lines up with a particular constellation. For example, during March, the Sun is somewhere between Pisces and Aquarius, and does not reach Aries until about the third week of April. The path of the Sun as determined by the boundaries of astronomical constellations[1] and the common zodiac dates are listed in table 7.2 for comparison[2].

Note that the Sun crosses through the constellation of Scorpius in less than 9 days and between 30 November and 18 December the Sun is in the constellation of

---

[1] The constellation boundaries are defined by the IAU see https://www.iau.org/public/themes/constellations/.
[2] The number of days listed in the third column is rounded off to the nearest integer.

Table 7.2. Astronomical and astrological constellations.

Name	Astronomical	Days	Zodiac	Days
Aries	19 Apr–14 May	26	21 Mar–19 Apr	30
Taurus	14 May–21 Jun	38	20 Apr–20 May	31
Gemini	21 Jun–21 Jul	29	21 May–20 Jun	31
Cancer	21 Jul–11 Aug	21	21 Jun–22 Jul	32
Leo	11 Aug–17 Sep	37	23 Jul–22 Aug	31
Virgo	17 Sep–31 Oct	45	23 Aug–22 Sep	31
Libra	31 Oct–21 Nov	21	23 Sep–22 Oct	30
Scorpius	21 Nov–30 Nov	8	23 Oct–21 Nov	30
Ophiuchus	30 Nov–18 Dec	18	Not Included	
Sagittarius	18 Dec–21 Jan	34	22 Nov–21 Dec	30
Capricornus	21 Jan–17 Feb	27	22 Dec–19 Jan	29
Aquarius	17 Feb–12 Mar	24	20 Jan–18 Feb	30
Pisces	12 Mar–19 Apr	38	19 Feb–20 Mar	30

Ophiuchus (the Serpent Bearer). Often, this constellation is mistakenly referred to in popular media as 'a new constellation' or a newly 'discovered 13th constellation', etc.

The common zodiac designation lists Aries as the first constellation to coincide with the March equinox. Today we observe a shift of about one month. The March equinox occurs when the Sun is in Pisces. The shift is due to the precession of the Earth's spin axis, as discussed in chapter 2. Precession causes the equinoxes to shift with a repeat cycle of about 26 000 years[3]. Over that period, the RA of the March equinox completes a full turn of 24 h. It follows that the shift of the March equinox (and the RA of the boundaries of the constellation) is 24/26 000 or 0.92 h per 1000 years. In the past 2000 years, the constellation boundaries shifted by about 2 h, which is the average width of constellations in the zodiac. This explains part of the difference between the right and left parts of table 7.2.

7.3 Star names

There are several systems for naming stars, therefore the same star can be known by different names or designations. The brightest stars usually have their own *proper* names, for example Sirius, the brightest star in the sky. Sirius is a star in the constellation of Canis Major (Latin for Big Dog). Proper names are usually of Arabic origin (Rigel, Deneb, etc) or come from Greek mythology (Castor, Pollux, etc). In the *Bayer designation*, the star is identified by a letter and the name of the constellation (in the Latin genitive) to which the star belongs[4]. Bayer used the Greek alphabet (α, β, γ, etc) which has only 24 letters. Once the Greek alphabet is

---

[3] See the discussion in chapter 2.
[4] The Latin genitives of constellations are listed in table 7.3.

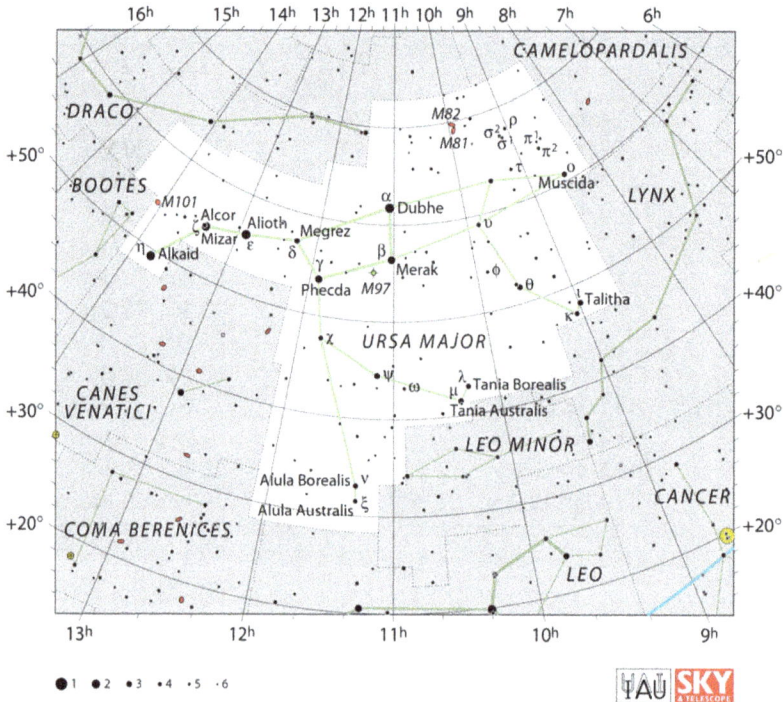

Figure 7.2. A map of Ursa Major. The Big Dipper is an asterism on the left (stars α, β, γ, ...η). Credit: International Astronomical Union, and *Sky and Telescope.*

exhausted, the Bayer system uses letters of the Latin alphabet and numbers. Letters from R to Z and double letters are used to indicate stars of variable brightness, for example T-Tauri (in Taurus), RR Lyrae (in Lyra).

The Bayer system generally (but not always) lists the brightest star in the constellation as the α (alpha) star. Thus Sirius is α Canis Majoris (*Majoris* is the genitive of *Major*). An exception to the rule occurs in the constellation of Orion. The brightest star in Orion is Rigel and Betelgeuse is the second in brightness. Contrary to the rule, in the Bayer designation α *Orionis* refers to Betelgeuse and β *Orionis* refers to Rigel. The brightness order of the stars does not always follow the order of the alphabet, instead at times they seem to follow the sequence of the constellation stick-figure. Figure 7.2 shows the constellation of Ursa Major (the Big Bear). The size of the dots representing the stars indicate the brightness of the star. A star magnitude[5] key is shown at the bottom. The Big Dipper is an asterism in the constellation of Ursa Major. The lettering of the stars and proper names are shown in figure 7.2. For example, the star Mergez is δ Ursae Majoris. The δ (delta) is the fourth letter of the Greek alphabet, but obviously Megrez is not the fourth brightest.

---

[5] The magnitude system is described in appendix D.

This is an example where the lettering does not follow the brightness, but the sequence in the stick-figure.

Other names refer to specific star catalogs. For example the HD and HDE (Henry Draper and Henry Draper Extension) is a list of about 300 000 stars. The most recent is the compilation from the Hipparcos and Tycho astrometry missions (http://www.rssd.esa.int/index.php?page=Overview&project=HIPPARCOS). The listing contains about 100 000 entries of nearby stars with accurate distance measurements.

In summary, a star may have different names depending on the system. Thus Arcturus (proper name) is also called α Bootis (Bayer designation), HD 124897 (Henry Draper catalog) and HIP 69673 (Hipparcos–Tycho catalog).

Of particular interest to observational astronomy is the Messier catalog. The catalog has 110 objects, including clusters of stars, remnants of dead stars and galaxies. Some of the Messier objects are visible to the naked eye. Some of the objects listed are fascinating when viewed using binoculars or a small telescope. The objects are named with an M followed by a number. For example M31 is the Great Galaxy in Andromeda.

7.4 Which constellations can be seen from a given location?

Not all constellations are visible from all locations on Earth and this includes the constellations of the zodiac. To determine which constellations are visible from a given location, one needs to compare the declination of the constellation to the latitude of the observer, as described in chapter 3. As a quick guide, from latitudes 45° all constellations with declinations less than −45° (e.g. −70°) are not visible. All constellations with declinations larger than 45° are visible all year round. At the Equator, all constellations are visible, although those with declinations above 75° and less than −75° may not be high enough above the horizon. From latitudes −45° all constellations with declinations less than −45° are visible all year round. All constellations with declinations larger than 45° are not visible.

In terms of the season when constellations are in the night sky, we can use the RA and follow the procedure described in chapter 3. For example, the RA of the Sun is about 0 h in late March to early April. During that time, all the constellations with RA near 12 h are high in the night sky around midnight (assuming of course that the constellations are visible from the observer's latitude). In late April to early May, the Sun's RA is about 2 h. During that time, all constellations with RA about 14 h are visible in the night sky around midnight, and so on[6].

Table 7.3 lists the astronomical constellations, with approximate RA and declination values. The map of each constellation can be downloaded from the official IAU page by clicking on the constellation name in the first column.

---

[6] See the discussion in chapter 3.

Table 7.3. A list of constellations. Adapted from https://www.iau.org/public/themes/constellations/.

Name	Genitive	Meaning	RA (h)	Dec (°)
Andromeda	Andromedae	Andromeda	1	40
Antlia	Antliae	Air Pump	10	−35
Apus	Apodis	Bird of Paradise	16	−75
Aquarius	Aquarii	Water Bearer	23	−10
Aquila	Aquilae	Eagle	20	5
Ara	Arae	Altar	17	−55
Aries	Arietis	Ram	3	20
Auriga	Aurigae	Charioteer	6	40
Bootes	Bootis	Herdsman	15	30
Caelum	Caeli	Chisel	5	−40
Camelopardalis	Camelopardalis	Giraffe	6	70
Cancer	Cancri	Crab	9	20
Canes Venatici	Canun Venaticorum	Hunting Dogs	13	40
Canis Major	Canis Majoris	Big Dog	7	−20
Canis Minor	Canis Minoris	Little Dog	8	5
Capricornus	Capricorni	Sea Goat	21	−20
Carina	Carinae	Keel	9	−60
Cassiopeia	Cassiopeiae	Cassiopeia	1	60
Centaurus	Centauri	Centaur	13	−45
Cepheus	Cephei	Cepheus	22	70
Cetus	Ceti	Sea Monster	2	−10
Chamaleon	Chamaleontis	Chameleon	11	−80
Circinus	Circini	Compass	15	−60
Columba	Columbae	Dove	6	−35
Coma Berenices	Comae Berenices	Berenice's Hair	13	20
Corona Australis	Coronae Australis	Southern Crown	19	−40
Corona Borealis	Coronae Borealis	Northern Crown	16	30
Corvus	Corvi	Crow	12	−20
Crater	Crateris	Cup	11	−15
Crux	Crucis	Southern Cross	12	−60
Cygnus	Cygni	Swan	21	40
Delphinus	Delphini	Dolphin	21	15
Dorado	Doradus	Swordfish	5	−65
Draco	Draconis	Dragon	17	60
Equuleus	Equulei	Little Horse	21	5
Eridanus	Eridani	Eridanus (a river)	4	−20
Fornax	Fornacis	Furnace	3	−30
Gemini	Geminorum	Twins	7	20
Grus	Gruis	Crane	23	−45
Hercules	Herculis	Hercules	17	30
Horologium	Horologii	Clock	3	−55
Hydra	Hydrae	A female sea snake	10	−20

(*Continued*)

Table 7.3. (*Continued*)

Name	Genitive	Meaning	RA (h)	Dec (°)
Hydrus	Hydri	A male water snake	2	−70
Indus	Indi	Indian	21	−55
Lacerta	Lacertae	Lizard	22.5	45
Leo	Leonis	Lion	11	15
Leo Minor	Leonis Minoris	Smaller Lion	10	35
Lepus	Leporis	Hare	6	−20
Libra	Librae	Scale	15	−15
Lupus	Lupi	Wolf	15.5	−45
Lynx	Lyncis	Lynx	8	45
Lyra	Lyrae	Lyre	19	35
Mensa	Mensae	Table	5	−80
Microscopium	Microscopii	Microscope	21	−35
Monoceros	Monocerotis	Unicorn	7	−5
Musca	Muscae	Fly	13	−70
Norma	Normae	Square	16	−50
Octans	Octantis	Octant	21	−85
Ophiucus	Ophiuchi	Serpent Bearer	17	0
Orion	Orionis	Orion	5.5	5
Pavo	Pavonis	Peacock	19.5	−65
Pegasus	Pegasi	Winged Horse	23	20
Perseus	Persei	Perseus	3.5	45
Phoenix	Phoenicis	Phoenix	1	−50
Pictor	Pictoris	Easel	5.5	−55
Pisces	Piscium	Fish	0.5	15
Pisces Austrinus	Pisces Austrini	Southern Fish	22	−30
Puppis	Puppis	Stern	8	−30
Pyxis	Pyxidis	Compass	9	−30
Reticulum	Reticuli	Reticle	4	−60
Sagitta	Sagittae	Arrow	19.5	18
Sagittarius	Sagittarii	Archer	19	−25
Scorpius	Scorpii	Scorpion	16.5	−35
Sculptor	Sculptoris	Sculptor	0.5	−30
Scutum	Scuti	Shield	19	−10
Serpens (Caput)	Serpentis	Serpent (head)	15.5	10
Serpens (Cauda)	Serpentis	Serpent (tail)	18	−12
Sextans	Sextantis	Sextant	10	0
Taurus	Tauri	Bull	4	15
Telescopium	Telescopii	Telescope	19	−50
Triangulum	Trianguli	Triangle	2	30
Triangulum Australe	Trianguli Australis	Southern Triangle	16	−65
Tucana	Tucanae	Toucan	0	−65
Ursa Major	Ursae Majoris	Great Bear	11	50
Ursa Minor	Ursae Minoris	Little Bear	15	75

Table 7.3. (*Continued*)

Name	Genitive	Meaning	RA (h)	Dec (°)
Vela	Velorum	Sails	9.5	−50
Virgo	Virginis	Virgin	13	0
Volans	Volantis	Flying Fish	8	−70
Vulpecula	Vulpeculae	Fox	20	25

7.5 Resources

A complete list of the Messier objects and a wealth of information can be found at http://messier.seds.org/.

The most complete and widely accessible list of star data is available from http://cds.u-strasbg.fr/.

Constellation maps and boundaries are available from https://www.iau.org/public/themes/constellations/.

Chapter 8

Star properties

Our understanding of star properties is based on analysis of the light we receive from them. By 'light' we mean not only visible radiation but, very importantly, radiation that the human eye cannot see, such as x-rays, ultraviolet light, micro-waves, radio waves, etc. In interstellar space, all these forms of radiation travel at the same speed, the speed of light, and collectively they form the *electromagnetic spectrum*.

Even at the speed of light, there is a significant lag between the time the radiation was emitted from the star and the time when the radiation is detected by Earth-based observers. For example, it takes about 8 min for light leaving the surface of the Sun to reach Earth. It takes over 4 years for light leaving Proxima Centauri, our nearest neighbor star, to reach us. In other words, what we see today is how Proxima Centauri was 4 years ago. In a way this time lag is an advantage because by looking farther into space astrophysicists are essentially looking back into the past and see stars and galaxies that are at different stages of their evolution. The images from nearby stars and galaxies represent how the Universe is 'now'. The images from more distant objects tell us how the Universe looked when it was younger. Understanding the evolution of stars and the Universe is one of the most exciting areas of research and is beyond the scope of this book. This chapter will provide a qualitative description of the principles of some of the methods used to measure some star properties.

8.1 The color and temperature of stars

Stars like the Sun obtain their energy from the hydrogen fusion that takes place deep in their interior. Fusion is a nuclear reaction that fuses four hydrogen nuclei to make a nucleus of helium, releasing energy and other particles in the process. Stars that obtain their energy from hydrogen fusion are called *main sequence stars*. Fusion can take place only in the core of the star, where the temperature is several million degrees and the pressure is about a trillion times larger than the atmospheric

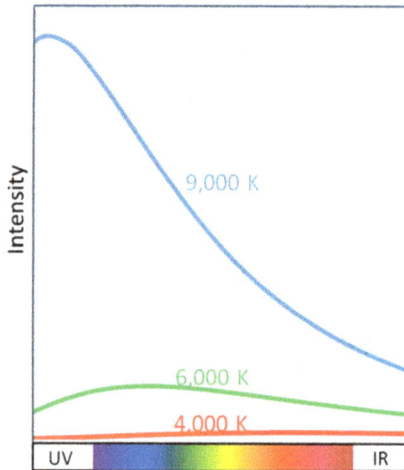

Figure 8.1. The continuous spectrum at three temperatures. Note that at higher temperatures the relative amount of blue increases.

pressure on Earth. These numbers are estimates, because the interior of a star cannot be observed directly.

What we perceive as the surface of a star is the opaque layer below the star's atmosphere, the so-called *photosphere*. The average temperature of the Sun's photosphere is about 5800 K[1]. The photosphere emits the sunlight (starlight) we see and the color of the light emitted depends on the temperature of the photosphere. From everyday experience we know that as a solid object becomes hot, it emits radiation (heat, which is infra-red light) and if the temperature of the object increases further it emits more and more heat, and may begin to emit visible light (it becomes red hot). The filament in an ordinary incandescent light bulb is even hotter than red hot, it is yellow to white hot.

Hot solids emit all colors of light, but at different proportions, depending on the temperature. What our eye perceives depends on which color dominates. If the light emitted contains red, green and blue in a certain balance, the human visual system perceives white light. If there is slight excess of red and less blue, we perceive yellow. As the excess of red increases, we see orange, orange-red, etc. The color distribution of the light emitted by a hot solid is the *continuous spectrum*. Figure 8.1 shows the continuous spectrum for three different temperatures. Note that as the temperature increases, the overall intensity is significantly increased. Note also that the increase is more dramatic in the blue.

If the temperature is low, below 4000 K, the emitted light is red. As the temperature is increased, the color becomes orange-red and at above 5000 K the color becomes golden-yellow, like the color of the Sun. Hotter stars (7500 to 10 000 K) look white, and even hotter stars look blue or blue-violet. As a general rule, the more red the star appears, the lower its surface temperature. The more blue the star

---

[1] The temperatures are given in the kelvin scale (K) which is equivalent to degrees centigrade (°C) + 273.

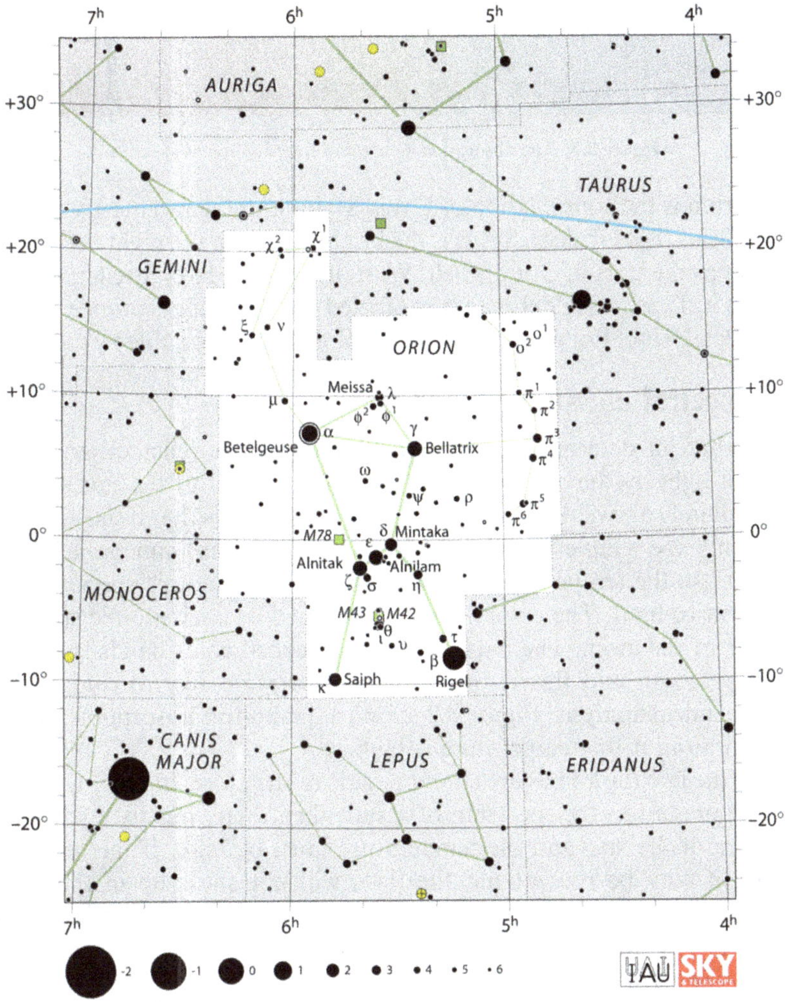

Figure 8.2. A star map of the constellation of Orion. Credit: International Astronomical Union and *Sky and Telescope*.

appears, the hotter its surface. Analysis of the continuous spectrum of the light emitted by a star provides one way of determining the surface temperature of the star. Note that there is a temperature range where green is in highest proportion. This is the temperature range where all the colors are in a certain balance such that to the human eye the star appears white, not green.

For the brighter stars the color is easily detectable. Figure 8.2 shows the star map in the constellation of Orion which is visible in the evening from mid-December to mid-April. The brightness of the stars is indicated by the size of the dots. In Orion the brightest stars are Rigel and Betelgeuse. Their color is distinctly noticeable. Rigel is blue-white (i.e. hot) and Betelgeuse is red (i.e. cool). The three stars Alnitak, Alnilam and Mintaka are distinctly blue, and together form *Orion's Belt*. These stars are very hot.

Figure 8.3. The absorption spectrum of the hydrogen atom.

East of Orion is the constellation of Canis Major and the brightest star in the constellation is Sirius (not labeled). Sirius is the brightest star in the night sky and is white, i.e. not as hot as the stars in Orion's belt. West of Orion is the constellation of Taurus. The brightest in Taurus is Aldebaran (not labeled) and its color is orange, i.e. cool, but hotter than Betelgeuse. For dimmer stars the color is not detectable by the human eye.

8.2 The spectral classification of stars

Another method of determining the surface temperature relies on analysis of the absorption of light by the *chromosphere*, the bottom layer of the star's atmosphere that is immediately above the *photosphere*. The basic idea is that if the temperature is very high, only the simplest atoms (hydrogen and helium) can be in a state that absorbs light. As the temperature becomes lower, more atomic species can be in a state that absorbs light. The color of light[2] absorbed by each atomic species is very *characteristic* of the atom. The pattern of the absorbed color bands is the so-called *absorption spectrum*, and this is the basis of *absorption spectroscopy*, a powerful method of chemical analysis. Figure 8.3 shows a qualitative absorption spectrum for the hydrogen atom if the temperature is high.

Note that the two dark lines are characteristic of the atom, in other words, if these two lines are present in the spectrum of a star, then there must be hydrogen in the chromosphere of the star and the temperature must be high. If the temperature is low, hydrogen may be present but the lines will not show up in the absorption spectrum. Therefore, the absorption spectrum contains information about the temperature of the chromosphere of the star. In terms of the absorption spectrum, stars are classified in seven main *spectral types*: O, B, A, F, G, K and M. The O type is the hottest and the M type is the coolest. Each type includes ten subdivisions, with 0 the hottest and 9 the coolest of the given spectral type. For example, for type A, we have A0, A1, A2, ..., A9, which is followed in order of decreasing temperature by F0, F1, ..., F9, then G0, G1, etc. In terms of spectral type, the Sun is a G2 star.

Table 8.1 lists the main spectral types with the corresponding temperature, color and a star representative of that spectral type.

8.3 Other information contained in the absorption spectra of stars

Analysis of the absorption spectrum can also provide information about the motion of stars. From everyday experience we know that the pitch of the siren we hear from fire trucks depends on their motion: if the siren is approaching the pitch becomes higher and if the siren is receding the pitch becomes lower. The same applies to light.

---

[2] The color is indicated by the so-called wavelength of light.

Table 8.1. The main spectral types of stars, with temperature range and color.

Spectral type	Temperature (K)	Color	Representative star
O	Above 25 000	Blue–violet	Na'ir al Saif, O9
B	10 000–25 000	Blue–white	Rigel, B8
A	7500–10 000	White	Sirius, A1
F	6000–7500	Yellow–white	Procyon, F5
G	5000–6000	Yellow	Sun, G2
K	3500–5000	Orange	Arcturus, K2
M	Below 3500	Red	Betelgeuse, M2

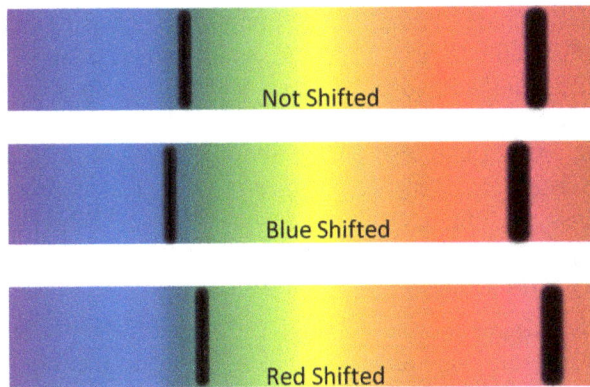

Figure 8.4. Qualitative absorption spectra of hydrogen showing the shift due to motion.

If the star is approaching, the absorption spectral lines move towards the blue end of the spectrum (*blue shift*) and if the star is receding the shift is towards the red end of the spectrum (*red shift*). A qualitative diagram of the shifted lines is given in figure 8.4. This shift of the spectral lines is called the *Doppler shift* and can be used to determine the speed of motion along the line of sight to the star. Note that the important quantity is the distance between star and observer: if the distance is becoming smaller, we have blue-shift, and it does not matter if the observer is moving towards the star or the star is moving towards the observer. In other words, there is only *relative motion* and there is no *absolute motion*. The Doppler shift is a very powerful method and was used to establish the *expansion of the Universe*, as well as the discovery of *exoplanets*.

In addition, the analysis of the absorption spectrum provides information about the pressure of the stellar atmosphere. If the pressure in the star's atmosphere is low, the absorption lines are sharp. If the pressure is high, then the absorption lines are broad. The two spectra in figure 8.5 show the effect of pressure on the width of the spectral lines.

The atmospheric pressure of a star is determined by the strength of gravity at the surface of the star. Stronger gravity will make the pressure higher and can result if the material in the star is densely packed. For stars, this means that the mass is

Figure 8.5. Qualitative absorption spectra of the hydrogen atom showing the effect of pressure on the width of the dark absorption lines. The dark absorption lines in the top spectrum indicate low pressure in the star's atmosphere. The dark lines are wider in the bottom spectrum indicating high pressure in the star's atmosphere.

packed in a smaller volume. Similarly, low pressure indicates low surface gravity, a lower density and therefore a larger volume, which means a larger radius. Thus by analyzing the width of the absorption lines, we can obtain some idea about the density and the size of the star itself.

The absorption spectra can also tell us about the magnetic field of the star, its rotation and also detect the existence of other stars or planets that may be orbiting the star. The discussion of these fascinating topics is beyond the scope of this book.

8.4 Luminosity and luminosity classes

From Earth, the Sun appears to be the brightest object in the sky because the Sun is close to Earth. Other stars may emit a lot more light than the Sun, but they appear dim because they are so far away. The situation is similar to light bulbs. For example, a 60 watt (W) light bulb may appear dazzlingly bright if it is located close by, while a 1000 W street light may appear quite dim if it is 1 km away. The terms *luminosity* or *intrinsic brightness* are used to characterize the rate of energy emission. The luminosity is a characteristic quantity of the star and therefore independent of the distance of the star, in the same way that a 100 W light bulb outputs 100 W independent of where it is.

Star luminosities are usually expressed using the Sun's luminosity as a unit. The luminosity of a star is determined by two quantities: the surface temperature and the size of the star. The higher the temperature, the more energy is emitted from each part of the surface. Also, the larger the star, the more surface there is. The luminosity or intrinsic brightness of stars is also expressed in terms of the *absolute magnitude*. The absolute magnitude scale is discussed in detail in appendix D.

The size–temperature combinations for different stars can be used to classify stars in terms of their luminosity. For example, *white dwarf* stars are much smaller than the Sun, but their surface temperature is much higher than the Sun. As a result, white dwarfs have very low luminosities. For example, Sirius, the brightest star in the sky, has a companion star that is a white dwarf, Sirius B. Sirius B is about the size of the Earth, but its surface temperature is over 20 000 K compared to the Sun's 5800 K. In spite of its high temperature, Sirius B is 10 000 dimmer than the Sun.

In contrast to white dwarfs, *red giant* stars have a lower temperature than the Sun, but are much larger. The result is that red giants have higher luminosities.

For example, Aldebaran, the brightest star in the constellation of Taurus, is a red giant. Its surface temperature is about 4000 K but its radius is over 40 times larger than the Sun's radius. As a result of its larger size, Aldebaran is over 500 times more luminous than the Sun.

Although white dwarfs are smaller than the Sun in size, this does not mean that the mass of the white dwarfs is much smaller than the Sun's mass. Actually, the mass of a white dwarf can be larger than the mass of the Sun. The difference is that the mass of a white dwarf is packed in a smaller volume, more densely. In fact the material that makes the white dwarfs is so dense that on Earth a spoonful of this material would weigh 1 ton! The reverse is true for red giants. They can be 10 times larger than the Sun, but have less mass than the Sun. In red giant stars the material is loosely packed. The difference in the star's density, and therefore atmospheric pressure, can be demonstrated by the width of the dark lines in the absorption spectrum as discussed in section 8.3. Red giants have narrower absorption spectral lines (which means lower pressure) while white dwarfs have broader spectral lines (which means higher pressure). Red giants and white dwarfs are stages in the evolution of main sequence stars (see section 8.1) after they have depleted the hydrogen in their core.

Depending on the luminosity, stars are classified in five luminosity classes. The luminosity classes are indicated by a roman numeral from I to V.

Luminosity classes:
   I. Supergiants
   II. Bright giants
   III. Giants
   IV. Subgiants
   V. Main sequence stars or dwarf stars

In luminosity class V, the term 'dwarf' does not include white dwarfs. White dwarfs are designated by the symbol D. This classification has been extended to include more classes, for example, class 0 hypergiants to indicate stars more luminous than class I, and class I is subdivided into Ia for bright supergiants and Ib for supergiants.

A star is completely characterized by its spectral type and its luminosity class. For example the Sun is a G2V (V means main sequence); Sirius B is a DA2 (D means white dwarf); Arcturus is a K2III (III means giant).

The star information is summarized in the so-called HR diagram[3]. The various luminosity classes are outlined in figure 8.6. Stars at the left end of the diagram are the hottest stars. Stars on the right end of the diagram are the coolest stars. The vertical axis is the luminosity expressed in solar luminosities. Stars towards the top of the diagram are over 100 000 times more luminous than the Sun. Stars at the bottom have luminosities that are 100 or more times less luminous than the Sun. The background colors in the diagram represent the approximate color of each spectral class. As discussed earlier, the color of a star relates directly to the surface temperature of the star.

---

[3] The HR diagram is named after Ejnar Hertzsprung and Henry Norris Russell.

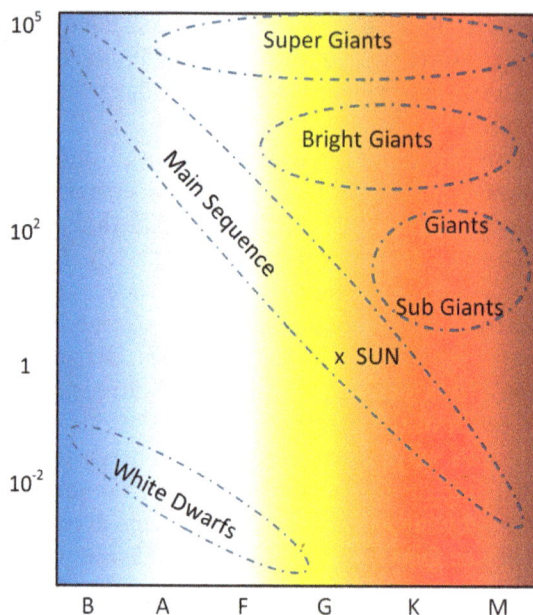

Figure 8.6. A qualitative HR diagram. The vertical axis indicates luminosity in terms of the Sun's luminosity. The background indicates color of the star. The axes of HR diagrams can be labeled in many ways. For example, the horizontal axis can indicate temperature instead of the spectral class[4].

8.5 The apparent brightness and the inverse square law

The apparent brightness, or simply *brightness*, refers to the amount of light we receive from a star. What determines the apparent brightness of a star is its distance from the observer and the star's intrinsic brightness or luminosity. A star may look bright because it emits lots of light, or because it is close to us, or both. The energy radiated by a star (which is what the luminosity tells us) spreads over larger and larger areas in all directions the farther it travels from the star. Therefore, the amount received by the observer on Earth depends on the rate the energy is emitted by the star, but also on the distance of the star from Earth. The dependence on distance follows the *inverse square law*, which states that the apparent brightness of a star depends on the inverse of the distance squared. Consider for example two identical stars (i.e. two stars of the same luminosity). If one of the stars is 10 times farther, it will not appear 10 times dimmer; instead it will appear 100 times dimmer, i.e. the brightness changes by a factor of $1/10^2$. The apparent brightness of stars is usually expressed in terms of the *apparent magnitude*. The apparent magnitude scale is discussed in detail in appendix D.

8.6 The size of stars

When viewed through telescopes, stars always appear as points. The disk seen in images is a spread of the light by diffraction (see chapter 9) and does not represent

---

[4] The scientific notation $10^2 = 100$, $10^3 = 1000$ and $10^{-2} = 1/100$ etc.

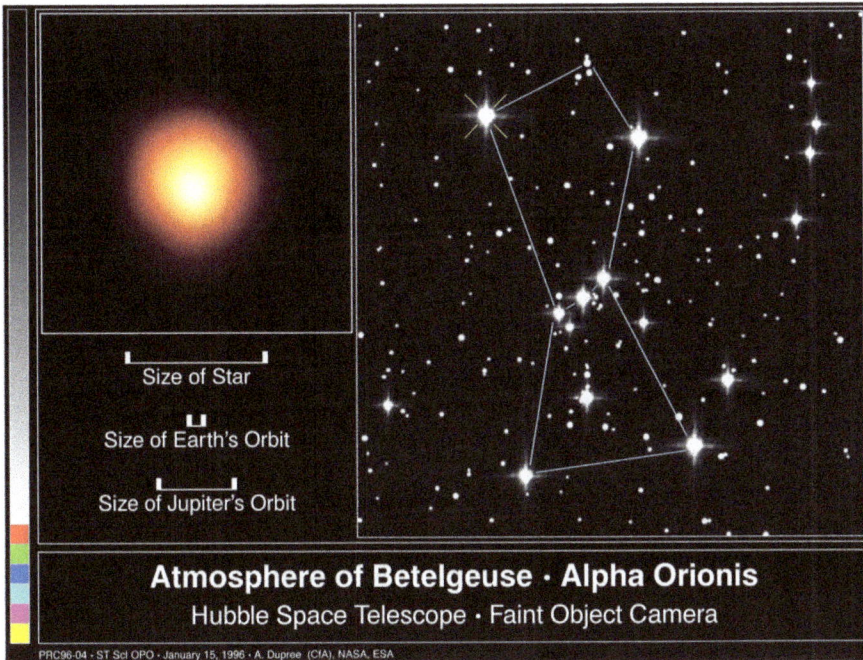

Figure 8.7. A resolved image of Betelgeuse, obtained by the Hubble Space Telescope. Betelgeuse (or alpha Orionis) is the star at the top left of the stick figure of the constellation. Credit: A Dupree (CfA), NASA, ESA, http://apod.nasa.gov/apod/image/9804/betelgeuse_hst_big.jpg.

the actual size of the star. This is true even for the best telescopes. To date, the number of star images that represent the actual size of the star (so-called *resolved* images) is fewer than a dozen. The first star to have its image resolved was Betelgeuse in the constellation of Orion, shown in figure 8.7.

The size of unresolved stars can be determined indirectly from the luminosity. The luminosity of a star depends on the temperature and the size of the star. Knowing the temperature and the luminosity allows the estimate of a star's size. If two stars have the same temperature (same spectral type) then the more luminous star must have a larger radius.

For comparison we will consider two stars of spectral type M, Betelgeuse and Menkar (the brightest star in the constellation of Cetus). The fact that the two stars are of the same spectral type means that in the HR diagram of figure 8.6 the two stars lie along the same vertical. The luminosity class of Betelgeuse is I and that of Menkar is III. Therefore Betelguese has larger luminosity, i.e. is closer to the top of the HR diagram than Menkar, so the radius of Betelgeuse must be larger than that of Menkar. The Sun's radius is commonly used as a unit for star radii. A detailed analysis of the data shows that Betelgeuse is approximately 1000 while Menkar is only 90 solar radii. In summary, this comparison shows that the radius of the stars increases as we move vertically from the bottom of the HR diagram to the top of the diagram in figure 8.6.

Visual Astronomy
A guide to understanding the night sky
Panos Photinos

Chapter 9

Telescopes

The use of mirrors and lenses to create images has been known since ancient times. Although Galileo did not invent the telescope, as it is incorrectly stated occasionally, he is recognized as the first to use the telescope as a tool for astronomy. The primary function of a telescope is to create a real image of a celestial object. In a real image, the light from the object is focused on a surface, be it a screen or a photographic film. For example, when using a lens to set a piece of paper on fire, the light from the Sun is actually focused on the paper, and the spot seen (before the smoke) is a real image of the Sun. To be of any use, the image must be a coherent representation of the object, i.e. a faithful correspondence of what is in the image to what the object looks like. Additionally, the image must be bright enough to show all the parts of the object, large enough and clear enough to see the details. In short, *the important qualities of a telescope are*:
   a. The light gathering power.
   b. The resolution.
   c. The magnification.

These characteristics apply to many imaging systems. For example, a photographic camera must take bright pictures and record picture details that allow high resolution and magnification.

9.1 Types of telescopes

The image forming element of a telescope, the so-called *objective*, can be a converging lens or a concave mirror. The objective is the most important part of the telescope. If the telescope objective is a lens, the telescope is a *refractor*. If the telescope objective is a mirror, the telescope is a *reflector*. *Catadioptric* telescopes are reflectors, but also have a lens at the front end of the tube to correct the image and allow a wider field of view.

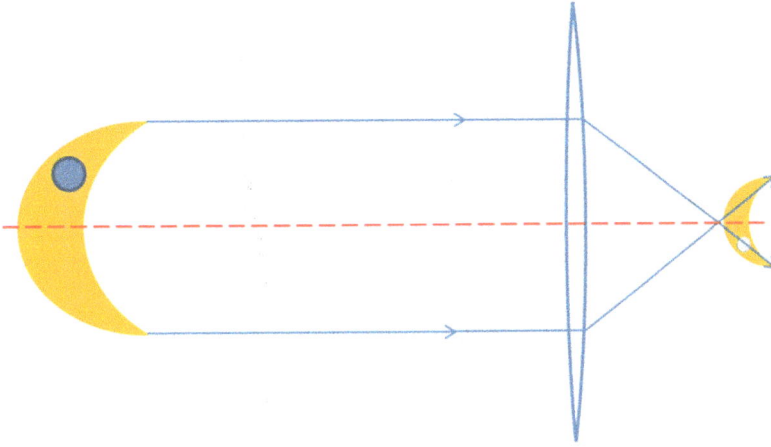

Figure 9.1. A ray diagram showing image formation by a converging lens. The object is on the left-hand side and the lens is in the middle. The rays cross the dotted axis at the *focal point* of the lens. The image is inverted and on the right hand side.

Figure 9.1 shows a diagram of the object (on the left) and image (on the right) formed by a converging lens. The real image is inverted and is located behind the lens, i.e. the lens is between the object and the image. The blue dot on the crescent is used as a marker to compare the orientation of the image relative to the direction of the object. As seen in the figure, the image is upside-down, because the light rays bend and cross before they form the image. For the same reason, if the object is three-dimensional, we would expect that what is above the plane of the figure, upon crossing the lens, will end up below the plane of the figure. This is indicated in the figure by changing the blue dot on the object into a white dot on the image. We also note that the parts of the object that are closest to the lens (the points of the crescent) become the farthest points on the image.

For astronomical work, the inverted image is of little consequence, other than the counterintuitive adjustments of the pointing direction. For example, if the image of the Moon appears at the top of the view through the telescope, the telescope is actually pointing towards the top of the Moon. And similarly, if the image appears on the right hand side of the view through the telescope, the telescope is pointing towards the right side of the object. However, caution is necessary when examining images (e.g. of the Moon or planets) as the posted images may or may not be inverted, and this piece of information is not always provided on the images.

Figure 9.2 shows a diagram of the object (left) and the image (middle) formed by a concave mirror.

The real image formed by concave mirrors is inverted, but note that the image is formed on the same side as the object, i.e. in front of the mirror. This poses a design problem for reflectors, which is solved by introducing a secondary mirror in front of the main, or *primary*, mirror. A flat secondary mirror can be used to deflect the light rays and focus the image to the side (Newtonian focus) of the telescope tube.

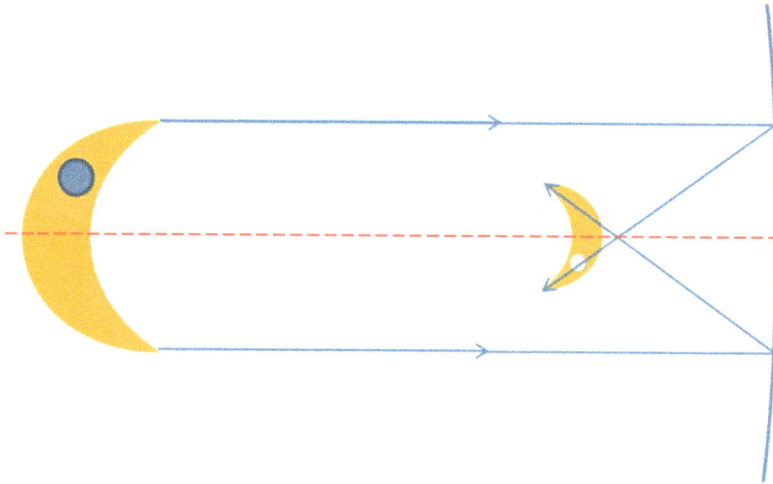

Figure 9.2. A ray diagram showing image formation by a concave mirror. The object is on the left-hand side, the mirror is on the right. The image forms in front of the mirror, and is inverted. The rays cross the dotted axis at the focal point of the mirror.

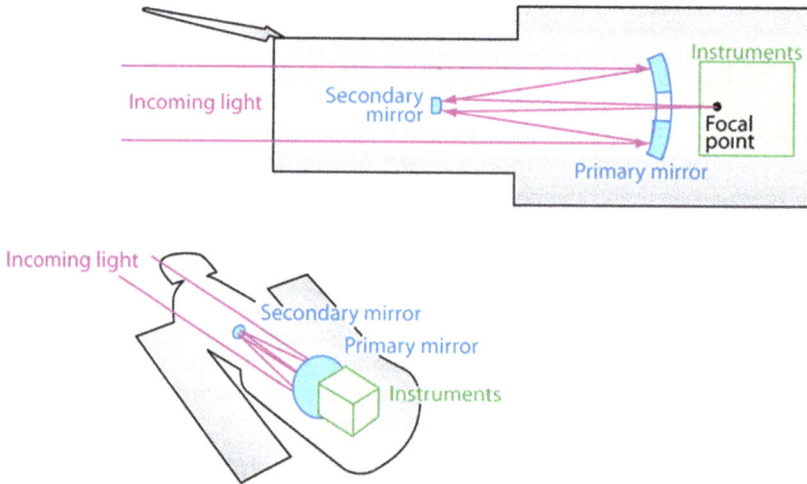

Figure 9.3. A schematic diagram of the Hubble Space Telescope. Credit: NASA and STScI, http://hubblesite. org/the_telescope/hubble_essentials/image.php?image=light-path.

A common alternative is to reflect the light rays back, through a hole in the primary mirror and focus the image behind the mirror (Cassegrain focus). In both cases the deflecting mirror (the so-called *secondary mirror*) and its support are in *front of the primary mirror* and they block some of the incoming light from the object. A schematic showing the path of light rays in the Hubble Space Telescope is shown in figure 9.3. The Hubble Telescope is a reflector and uses the Cassegrain design.

9.2 The focal length and the f/number

Lenses and mirrors are characterized by two numbers, the *aperture* and the *focal length*. For telescopes, the aperture is the diameter of the objective lens or mirror. The amount of light entering the telescope depends on the area of the aperture, therefore a larger aperture means more light gathering power. The focal length is the distance of the focal point from the center of the objective lens or the primary mirror (see figures 9.1 and 9.2). The image of an infinitely distant object forms at the focal point of the lens or mirror. Therefore, for telescopes, the focal length is essentially where the image of an infinitely distant object would form. For refractors, and for the Newtonian reflectors, the length of the tube is approximately the same as the focal length of the objective. For Cassegrain reflectors, the length of the tube is usually less than half the focal length, meaning that these telescopes are much shorter. Large reflectors in observatories do not have a tube.

It is common to give the relative aperture of the objective, which is the ratio of the focal length divided by the aperture. The relative aperture is usually referred to as the *f/number*. For the calculations it is important to have consistent units. Here we will use millimeters. For example, for a 114 mm objective (meaning an objective with diameter or aperture of 114 mm) and a focal length of 500 mm, the ratio is 500/114 = 4.3 which is specified as f/4.3 for brevity. These numbers are typical for a small reflector. Usually the f/number and the aperture are given. The focal length can be found by multiplying the aperture by the f/number. The f/number tells us how much light enters the telescope; a large f/number means less light is entering the telescope.

9.3 Magnification

In both reflector and refractor telescopes, the image formed by the objective can be viewed using a second lens, the eyepiece. The eyepiece has short focal length and it usually consists of a combination of lenses all in one piece. The magnification of the telescope is determined by the ratio of the focal length of the objective divided by the focal length of the eyepiece. For example, an eyepiece of focal length 25 mm, used with an objective of focal length 500 mm, will give magnification of 500/25 = 20×. The '×' meaning 'times'. If the eyepiece has a focal length of 4 mm, then the magnification is increased to 500/4 = 125×. As the focal length of the objective cannot be changed, different magnifications can be achieved by switching eyepieces. Telescopes should have at least two eyepieces, to give a minimal selection of magnifications.

Increasing the magnification is not always desirable. The field of view, which is what one sees through the eyepiece, decreases with increasing magnification. For example, when viewing four craters on the Moon's surface in our field of view, doubling the magnification will show a lot less, probably just one crater. In addition, the light coming from this one crater into the telescope will now be spread over the entire field of view, in other words, the image will be about 4 times dimmer. Finally, with increasing magnification, there comes a point where we lose sharpness: the image becomes fuzzy and further magnification is of no use. This point will be discussed next.

The useful magnification depends on the diameter D of the aperture of objective. If D is given in centimeters, then a rule of thumb is:

    maximum magnification: $16 \times D$
    minimum magnification: $1.6 \times D$.

When using inches, the rule is:

    maximum magnification: $40 \times D$
    minimum magnification: $4 \times D$.

For example, the useful range of magnification for an aperture of 10 cm (about 4 inches) will be from about 16× to 160×. It is of little value to buy an eyepiece that would give a magnification of 250× for this telescope.

9.4 Image resolution

When viewed through a telescope, stars appear as small disks. These are known as *Airy disks.* The brightness is highest at the center and diminishes towards the edges of the disk. The size of these disks does not represent the size of the stars and the disks become larger for telescopes with a smaller aperture. If two stars are very close together in the field of view, the disks may overlap and the two stars will no longer be distinguishable.

Suppose that we are observing two stars of similar brightness and the angle between the two stars as seen by the naked eye is symbolized by the Greek letter θ. If the angle is very small, the disks will overlap. Obviously, the resolving power is higher if the angle θ can be very small without the disks overlapping. Therefore, the resolving power is measured by the smallest angle between two stars that can be resolved as separate stars. This angle is the *angular resolution* and for visible light, under otherwise perfect conditions, is given in seconds of arc $(")$[1] by the formula

$$\theta = 14/D$$

where D is the aperture of the objective in centimeters. For example, if the aperture $D = 10$ cm, the angular resolution is $\theta = 14/10 = 1.4"$, meaning that the telescope can resolve stars that are $1.4"$ apart. A 100 cm objective will resolve stars that are $0.14"$ apart. A smaller angular resolution means better resolving power.

The limit of the angular resolution of a telescope is the result of the so-called *diffraction* of light. The dependence of the angular resolution on the aperture can be interpreted as follows. Light is diffracted when passing through an opening, for example a hole or an aperture. Light coming from the edges of the aperture combines to produce faint rings and fringes. If the edges are father apart (i.e. if the aperture is larger) the disks and fringes become fainter. Obviously having the edges farther apart means a larger aperture, or larger objective.

---

[1] See appendix A.

9.5 Refractors versus reflectors

As the light goes through the lens, different colors are bent by different amounts (this is called *dispersion*). The result is a rainbow-like effect, as observed in common glass prisms. Additional lenses are required to correct this dispersion of different colors. But more lenses means more optical surfaces to polish and more glass. The result is that refractors are heavier and more expensive. Mirrors have one optical surface, and since the light does not penetrate the surface, there is no dispersion. Reflectors are lighter and less expensive. On the down side, the mirror's reflecting surface (usually a coating of aluminum) deteriorates with time.

9.6 The effects of the atmosphere

The useful magnification range and the minimum angular resolution assume ideal components and also ideal observation conditions. Observing through the atmosphere introduces more limitations on the resolution. Air currents and temperature differences in the atmosphere along the line of sight produce their own spread of the image points and the amount of spread can be significant. In most places, the effects of the atmospheric conditions, known as *atmospheric seeing*, can introduce a spread of the image equivalent to 1–3″, and the effect becomes worse when viewing objects that are low on the horizon. Therefore, a telescope with an angular resolution smaller than 1″ cannot be used to its fullest capability in such an environment. The situation improves by going to higher altitudes where the air is thinner. This is the primary reason why large telescopes are located on top of high mountain peaks, or in space.

9.7 Using cameras with telescopes

Recording images with a telescope requires a sturdy tripod, a tracking system and a camera. The aperture of the telescope determines the light gathering power of a telescope. A larger aperture means more light enters the telescope. For visual observation through the telescope, the aperture is the deciding factor in determining the dimmest object that can be seen. If the aperture is small, faint stars will not be visible. For imaging, there is another alternative. The amount of light gathered to form an image can be accumulated over time. In other words we can use longer exposures to capture the image if we can keep the image from moving out of the field of view.

The first step is to eliminate any vibration of the telescope. This requires a rigid and sturdy tripod. The next step is to compensate for the motion of the celestial sphere. A tracking system uses one or two motors at the base of the telescope that continuously move the pointing direction of the telescope at a rate that compensates for the motion of stars. The Dobsonian telescopes (essentially reflector telescopes) are not equipped with a tracking mechanism and are not suitable for photography work.

For ordinary imaging purposes the charge coupled device (CCD) cameras are the most practical solution. The CCD cameras are large arrays of small detectors.

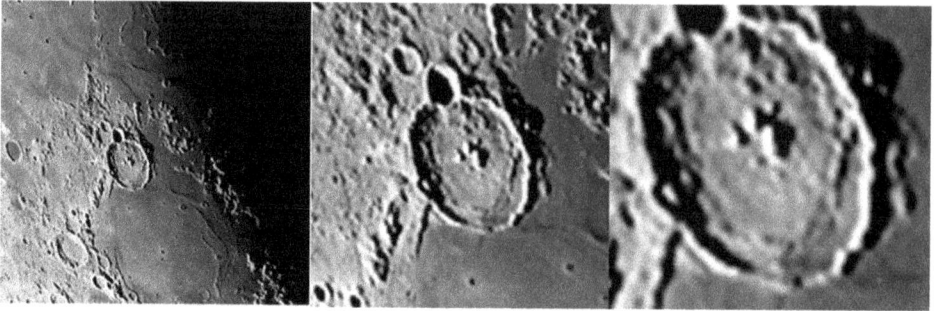

Figure 9.4. The resolution deteriorates as the magnification increases from left to right. The Moon's north pole is towards the top of the image. Credit for original image of crater Gassendi: ESO, http://www.eso.org/public/archives/images/publicationjpg/eso9903c.jpg.

Each detector records one part of the image (one picture element, or *pixel*). The recorded image therefore is a mosaic of pixels and the quality (the resolution) of the image can improve if the number of pixels is high. Low cost cameras for telescope use are available with 5 MP (i.e. 5 million pixels) CCDs and larger pixel numbers are becoming more affordable. The resolution of the image is ultimately determined by the angular resolution of the telescope and the atmospheric seeing. Therefore using more pixels does not necessarily improve the sharpness of the images.

The CCD cameras are connected to and controlled by laptop computers that directly digitize the captured image. The communication between the computer and CCD requires a 'driver' which should come with the CCD. The CCDs usually come with the software that allow further processing and editing of the images. To mount the camera, the eyepiece is removed and the camera is inserted in the tube. Therefore, in choosing a CCD, caution is required to match the size of the barrel of the camera and the size of the eyepiece of the telescope.

The magnification for a given telescope can be selected by using the proper eyepiece. Without the eyepiece, the size of the image seen by the camera is determined by the focal length of the objective, which is a characteristic of the telescope that cannot be changed. Some magnification can be achieved by using the so-called Barlow lens, which may come with the telescope. Further magnification can be achieved using software, which can be limited by pixelation. The images in figure 9.4 show a progression in magnification of a lunar crater. As the magnification is increased, the resolution deteriorates.

9.8 Finding the stars

Most modern telescopes have two motors that can control the direction of the tube in terms of altitude and azimuth[2]. Aligning the telescope and finding stars and

---

[2] See the altitude and azimuth coordinate system in chapter 3.

planets is all performed by a computerized controller that instructs the two motors to point the tube to the selected object. The user needs only to attach the telescope to a level tripod, enter information of the location (nearest city, country, local time and date) and finally point the telescope to three bright stars. The controller is then ready to find any objects (stars, planets, clusters, etc) that are included in the built-in database. Using a hand control, the user can normally scroll through large lists (tens of thousands) and select the objects to be seen.

Visual Astronomy
A guide to understanding the night sky
Panos Photinos

Appendix A

Measuring angles

The degree (°) is a unit of angle. A degree has 60 min of arc (′) and each minute has 60 s of arc (″). In symbols $1° = 60′$ and $1′ = 60″$. The terms seconds of arc, arc-seconds, etc, are used to distinguish angle units from the corresponding units of time. It follows from the above that $1° = 60 \times 60″ = 3600″$. A 90° angle is formed between two directions that are mutually perpendicular.

It is common to express angles in decimal form rather than sexagesimal form, i.e. using minutes and seconds. To convert angles in decimal form to sexagesimal, divide the minutes by 60 and the seconds by 3600. Add the results of these divisions to the degrees in the original angle. The sum is the angle in decimal form. For example, convert the angle 30° 20′ 40″ to decimal form. We will work with four decimal places. The number of degrees in the original angle is 30°. To this value we add the number of minutes (20) divided by 60 ($20/60 = 0.3333°$) and the seconds (40) divided by 3600 ($40/3600 = 0.0111°$). Adding all together:

$$30° + 0.3333° + 0.0111° = 30.3444°.$$

$$\text{Therefore, } 30° \ 20′ \ 40″ = 30.3444°.$$

To convert a decimal to sexagesimal, the whole part of the number is the number of degrees. Multiply the decimal part (right of the decimal point) by 60. The whole part of the result is the number of minutes (′). Next multiply the decimal part times 60. This is the number of seconds (″). For example, convert 18.2462° to sexagesimal. The number of degrees is the whole part, i.e. 18°. The decimal part is 0.2462. We multiply times 60, to obtain $0.2462 \times 60 = 14.772$. The whole part, 14, is the number of minutes. We multiply the decimal part, 0.772 times 60 to obtain $0.772 \times 60 = 46.32$ which is the number of seconds. Therefore, $18.2462° = 18° \ 14′ \ 46.32″$.

A.1 Angles in radians

Although the measurement of angles in degrees is more common, the measurement of angles in radians (which appears as an option in most calculators) is more suitable

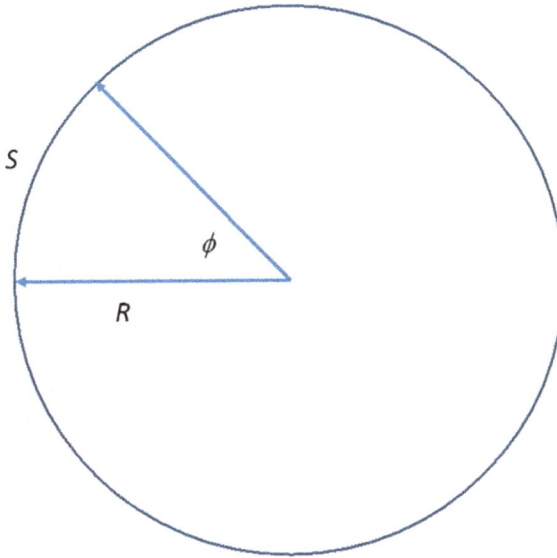

Figure A.1. The angle ϕ in radians is the length of the arc S divided by the radius of the circle R.

in some cases. The radian is a unit that derives from a more rigorous definition of the angle and is more helpful in understanding some important quantities in astronomy.

Figure A.1 shows a circle of radius R. The angle (ϕ) between the two lines can be measured as follows:

Angle (in radians) is equal to the length of arc S divided by radius of circle R. In symbols:

$$\phi = S/R \text{ (radians)}. \tag{A.1}$$

For example, if the angle between the lines is 90° then we have a quarter of a circle. To obtain S the length of the arc of a quarter circle we divide the circumference of the circle ($2\pi R$) by 4, or

$$S = (2\pi R)/4 = \pi R/2.^{[1]}$$

Following the definition above, the angle corresponding to the quarter circle (90°) is

$$\phi = S/R = \pi R/(2R) = (\pi/2) \text{ radians}.$$

Therefore, 90° equals ($\pi/2$) radians or about 1.57 radians. The complete circle (i.e. the 360°) corresponds to 2π radians.

From the above it follows that multiplying the angle in degrees by π and dividing by 180 gives the angle in radians. For example, to convert 30° degrees to radians:

$$30° \times \pi/180 = 0.5236 \text{ radians}.$$

---

[1] Here π is the familiar number 3.14...

To convert radians to degrees, multiply the angle in radians times 180 and divide by π. For example, to convert 1.5 radians to degrees:

$$1.5 \times 180/\pi = 85.9°.$$

A.2 Angular size

Figures A.2(a) and (b) show two different angles. The angle is decreasing from A.2(a) to (b). Assume that the observer is at the center of the circle, observing the object represented by the green line. We will call the length of the green line the *size of the object*. The center of the object is at a distance d from the observer. The size of the object is obviously smaller than the length of the arc S and the distance of the object d is smaller than the radius of the circle R. As the angle becomes smaller, the size of the object approaches the length of the arc S and the distance of the object approaches the radius R. Therefore, we can rewrite (A.1) above as:

$$\phi = (\text{size of object})/(\text{distance of object}) \qquad (A.2)$$

and the ratio is the definition of the angle between the two lines in radians. This angle is also referred to as the *angular size* of the object. The method is more accurate if the angles are small, i.e. when the size of the object is small compared to the distance, as is the case for astronomical objects.

Example. The angular size of the Moon

The diameter of the Moon is 3474 km and the average distance to Earth is 382 500 km. As the distance is much larger than the actual size of the Moon, we can apply (A.2) above to find the angular size of the Moon in radians, thus:

$$\phi = (3474)/(382\,500) = 0.00908 \text{ radians.}$$

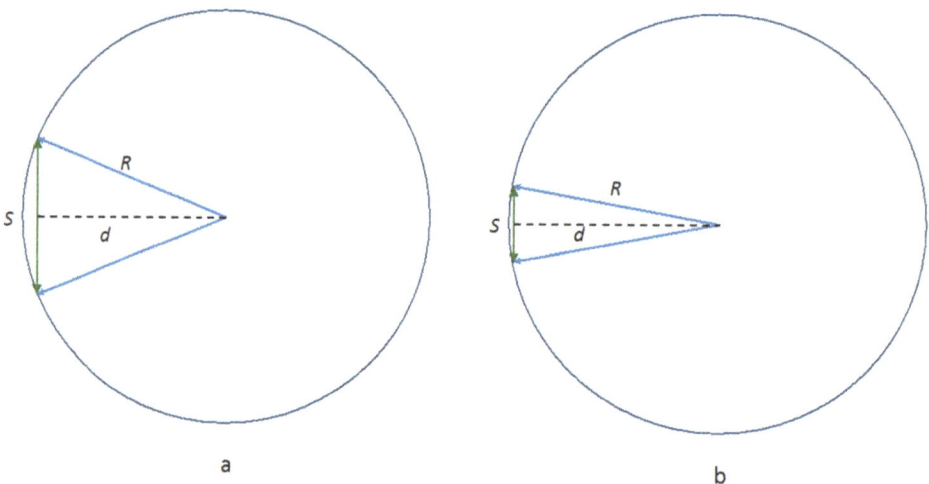

Figure A.2. The angular size of an object as seen by an observer located at the center of the circle. See the text for details.

As above, we convert to degrees by multiplying by 180 and dividing by π to find:

$$\phi = 0.52°.$$

At its closest, the distance of the Moon is 357 000 km and the angular size is

$$\phi = (3474)/(357\,000) = 0.00973 \text{ radians or } \phi = 0.56°,$$

about 7% more than average. When the full Moon phase occurs at about the near distance, the disk appears about 7% larger and is referred to as the *super Moon*.

A.3 The small angle formula

The angle calculated by (A.2) is in radians. We can convert the radians into degrees by multiplying by 180 and dividing by π, and further convert degrees to arcseconds by multiplying times 3600, and since $180 \times 3600/\pi = 206\,265$ we have:

$$\phi = 206\,265 \times (\text{size})/(\text{distance})$$

in arcseconds (″). The above equation is called the *small angle formula* and is important because it allows the calculation of size or distance. For example, the brightest star Sirius has an orbiting companion star, Sirius B. As observed from Earth, the average angular size of the orbit is about 7.5″ and the average distance from Earth is 8.6 light years (ly). We can estimate the average size of the orbit by rearranging the terms in the small angle formula. Thus

$$(\text{size}) = (\text{distance}) \times (\phi)/206\,265 = 8.6 \times 7.5/206\,265 = 0.00031 \text{ ly}.$$

There are 63 271 astronomical units (AU) in one light year, therefore the size of the orbit of the Sirius binary system is 19.6 AU (= $0.00031 \times 63\,271$), i.e. about 20 times larger than the average size of the Earth's orbit, i.e. slightly larger than the orbit of Uranus (see table 5.2).

The small angle formula is used to define the parsec. As the Earth orbits the Sun, the nearby stars appear to shift with respect to the background of distant stars[2]. The *parallax* is half the angle of shift between two observations from Earth made 6 months apart. We can rearrange the small angle formula to read

$$(\text{distance}) = 206\,265 \times (\text{size})/(\phi).$$

In this case, the angle refers to the parallax in arcseconds and the size is the semi-major axis of the Earth's orbit, i.e. 1 AU. Therefore

$$(\text{distance}) = 206\,265 \times (1\,\text{AU})/(\phi).$$

---

[2] See appendix B for details.

The parsec is defined as a distance equal to 206 265 AU. In terms of parsecs, the small angle formula reads

$$(\text{distance}) = 1/(\phi)$$

where the angle is in arcseconds.

A.4 A practical way of measuring angles

The precise measurement of angles is of prime importance in astronomy and requires accurate instruments called *goniometers*. For stargazing purposes, a simple yet very practical method can be used, as described below. For most people, the ratio of the width of their hand to the length of their arm is nearly constant. There are several hand positions that can be used: fist, fingers fully splayed, one finger, etc. Each of these positions measures a different angular distance. By using these hand positions, angular distance/size measurements can be made that are adequate for most naked-eye observations. The basic procedure is as follows.

Hold your arm at full length, close one eye and sight along the arm with the other eye. Then, based on how you hold your hand, you can measure different angles.

An angle of 20° corresponds to the span of your open hand at arm's length (see figure A.3). The width of the fountain spans about 20° or, equivalently, the *angular size* of the bowl is 20°. Note that the (linear or actual) *size* of the bowl is about 1.2 m (we are referring to *length* in this case!) but the angular size is an *angle*. If the distance of the object from the observer is increased, the angular size decreases, as shown in the following two figures.

Figure A.3. An open hand at arm's length. The angular size of the fountain is about 20°.

An angle of 10° corresponds to the width your full fist at arm's length (see figure A.4).

An angle of 1° corresponds to the width of the tip of the little finger at arm's length (see figure A.5).

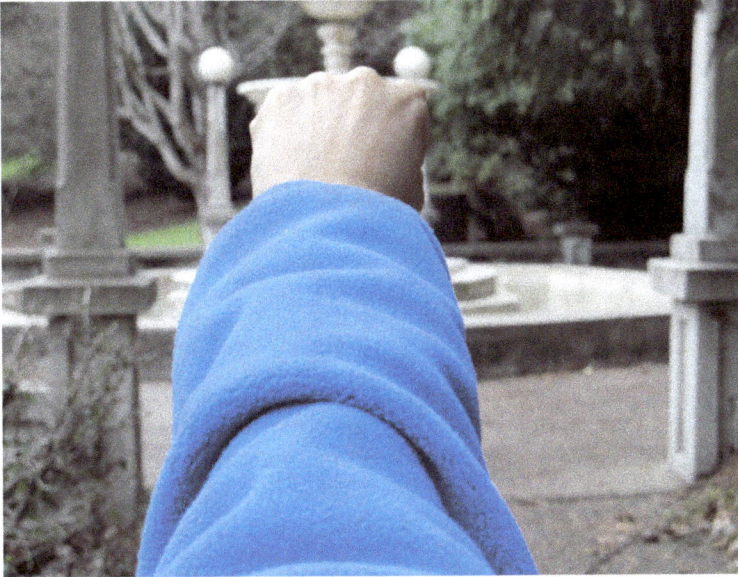

Figure A.4. A fist at arm's length. The angular size of the fountain is about 10°.

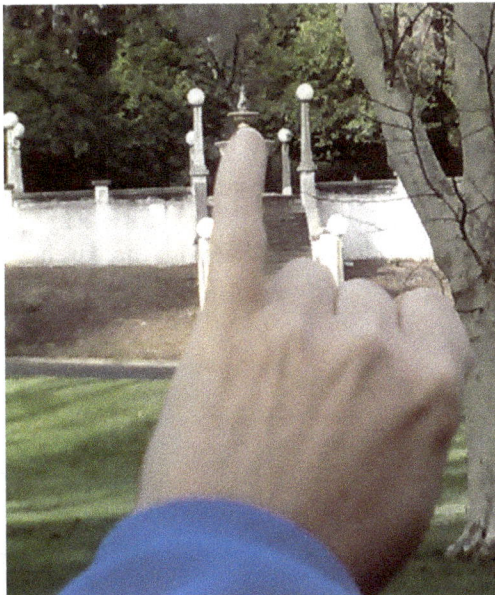

Figure A.5. The tip of the little finger at arm's length. The angular size of the fountain is slightly over 1°.

IOP Concise Physics

Visual Astronomy
A guide to understanding the night sky
Panos Photinos

Appendix B

Measuring distance in astronomy

Measuring the distance of celestial objects is a fundamental and difficult task. For nearby stars (less than about 500 light years (ly) away) the method of parallax is used. For more distant objects, more indirect methods are used, which are ultimately based on the parallax method.

B.1 The parallactic shift

The essential idea is that of perspective: the apparent shift of an object's position against a distant background, when the object is observed from different positions. Figure B.1 illustrates the concept. The two photographs were taken from the same distance from the signpost. Figure B.1(*a*) was taken first. The camera was moved 4 m to the left and figure B.1(*b*) was taken. Note that the signpost shifted from left to right relative to the background. The two photographs (observations) were taken from different positions and the result is a *parallactic shift*, or *parallax* for short. There is an inverse relation between the magnitude of the parallactic shift and the distance of the object (signpost) to the observer (camera).

The distance between the two observation points is called the *baseline*. In this example the camera was moved by 4 m, therefore the baseline was 4 m.

Figure B.1. Photographs of a signpost from different positions. See the text for details.

Using simple trigonometry one can find the distance of the object if the baseline and the shift are known. The parallax method can be applied to measure the distance of celestial objects. The method is the basis of the High Precision Parallax Collecting Satellite (Hipparcos) mission and also of the Panoramic Survey Telescope & Rapid Response System (Pan-STARRS) in Hawaii. Pan-STARRS can detect approaching comets and asteroids. Pan-STARRS uses four charge coupled device cameras of 1.4 billion pixels (gigapixels) each. The four cameras are mounted on a single telescope to provide four (rather than two) observation perspectives. In what follows we will discuss the application of the parallax method to measuring star distances and the distance to an asteroid.

B.2 A method for measuring the distances of stars

As the Earth moves around the Sun, our position in space is constantly changing. Figure B.2 shows the Earth in two different positions, in January and July.

As shown in figure B.2, our line of sight to a nearby star appears at a different point among the distant stars. The angle between the two lines of sight can be measured by making two observations 6 months apart, for instance in January and July. The baseline is 2 AU. Here we will work with half the angle and half the

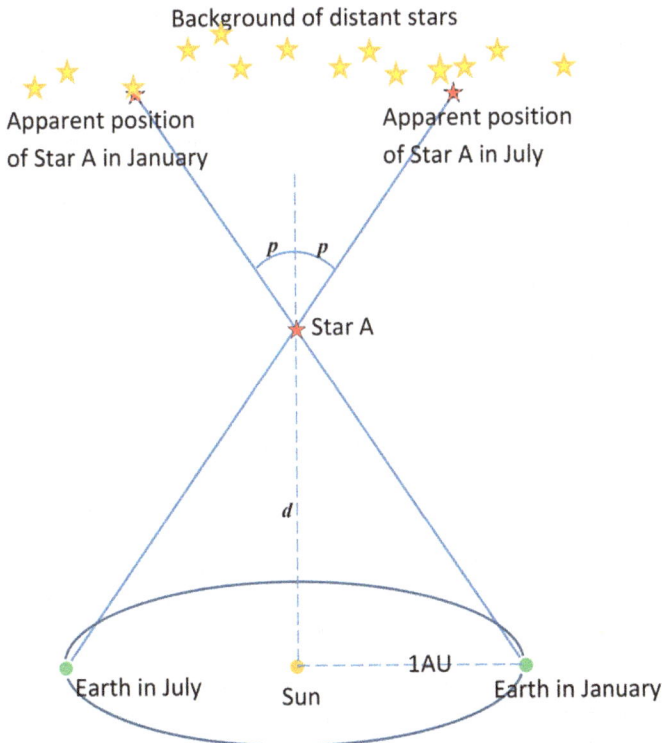

Figure B.2. The shift in the apparent position of a star, as seen from Earth in January and in July. Drawing not to scale.

baseline. The angle p is the *parallax angle*. The distance to the star d in astronomical units (AU) is related to the angle p (half the shift) and the Earth–Sun distance (which is 1 AU) by:

$$d = 206\,265/p$$

where the parallax angle p is measured in seconds of arc[1].

B.3 The parsec

For star distances it is convenient to define a unit of length equal to 206 265 AU. This unit is the parsec (pc).

$$1\text{ pc} = 206\,265\text{ AU}$$

and the distance formula becomes

$$d = 1/p$$

where the angle p is expressed in seconds of arc and the distance d in parsecs. As expected, the parallax angle is inversely related to the distance, i.e. the parallax is smaller for the more distant stars. The star Pollux (one of the twins of Gemini) has a parallax of about 0.1 arcseconds. Using the formula, we find that the distance of Pollux is 10 (= 1/0.1) pc. The star Acrux, the brightest in the Southern Cross, has parallax angle of about 0.01 arcseconds. Using the formula we find the distance of Acrux to be 100 (= 1/0.01) pc. Stars at distance of 1000 pc would have a parallax angle of 0.001 arcseconds, which is a very small angle (about 10 000 times smaller than a human hair held at arm's length) and beyond present measurement capabilities.

The parsec and the light year are commonly used units to express star distances. One can convert parsecs to light years using the formula:

$$1\text{ pc} = 3.26\text{ ly.}$$

B.4 Measuring the distance of nearby objects

The parallax method can be used for measuring the distance of nearby objects, such as asteroids, comets and the path of meteorites[2]. The principle of the method is the same as for stars: two observations are made from two different locations and the distance is determined by knowing the baseline and measuring the parallax angle. Nearby objects shift position in the sky faster than stars. Therefore, the two observations must be done simultaneously.

For nearby objects we can use two simultaneous observations from two different parts of the Earth, as shown in figure B.3.

---

[1] The number 206 265 is used to convert the radians to seconds of arc. See appendix A.
[2] The distances of the planets and the Moon are currently measured using radar techniques.

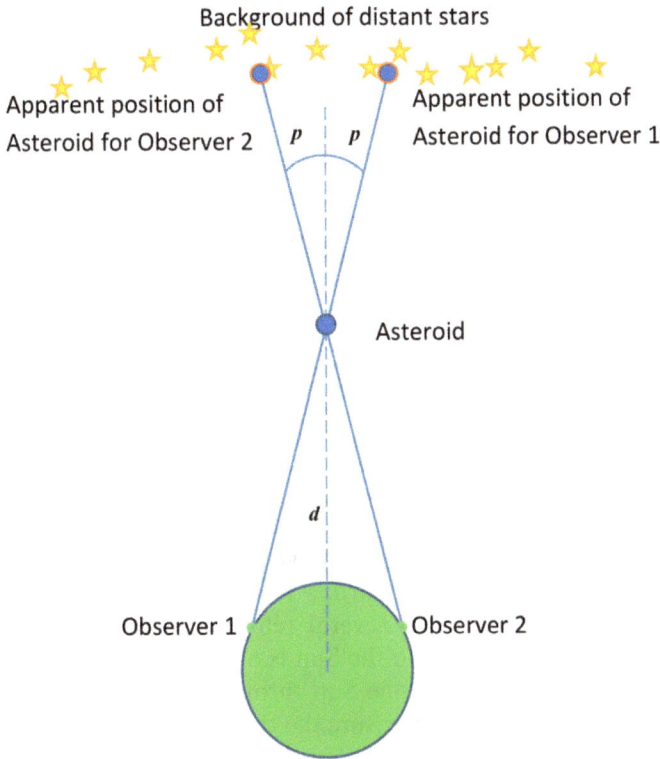

Figure B.3. The shift of a nearby object (e.g. an asteroid) as seen simultaneously by two observers on Earth. Distances not to scale.

Knowing the (linear) distance between the two observers, we can calculate the distance of the object d. The precision of the method increases if the distance between observers is large, e.g. 500 km. To determine the height of meteorites, photographs taken by two observers separated by a distance of approximately 20 km are adequate.

Visual Astronomy
A guide to understanding the night sky
Panos Photinos

Appendix C

Time keeping

The repeating cycles of celestial motions have been used to mark time and make calendars since ancient times. The Earth's rotation provides one such measure, although seasonal variations made several refinements necessary. One complete rotation of the Earth with respect to the Sun is a solar day. It can be measured by marking two successive transits of the Sun through the meridian of the observer. An ordinary clock can be used to measure the time interval between two successive noon positions as marked by a sundial. This is known as the *apparent solar* day.

The length of the apparent solar day is not the same from day to day. This is primarily the result of the tilt of the Earth's rotation axis with respect to its orbit and the elliptical shape of the orbit. The apparent solar day in December could be as much as 50 s longer than in September[1]. The cumulative effect of a succession of shorter days could lead to a difference of several minutes between the Sun transit as marked by the sundial and noon as indicated by an ordinary clock.

The time as measured by the 24 h clock is referred to as the *mean solar time* and the mean solar day starts at midnight. All locations with the same longitudes have the same mean solar time. This is different from *time zones*, as discussed below. The numerical difference:

'apparent solar time' – 'mean solar time'

is referred to as the *equation of time* and is shown in figure C.1.

To understand the significance of the numbers in the vertical axis of figure C.1, let us assume an observer at the prime meridian at Greenwich, using standard time (as opposed to daylight savings time; explained below). For January 1, the number of minutes indicated on the vertical axis of the figure is −4, meaning that the Sun will cross the meridian at 12:04 according to the clock. For the end of October, the number of minutes is 16, meaning that the Sun will cross the meridian 16 min

---

[1] The apparent solar day should not be confused with the daylight hours.

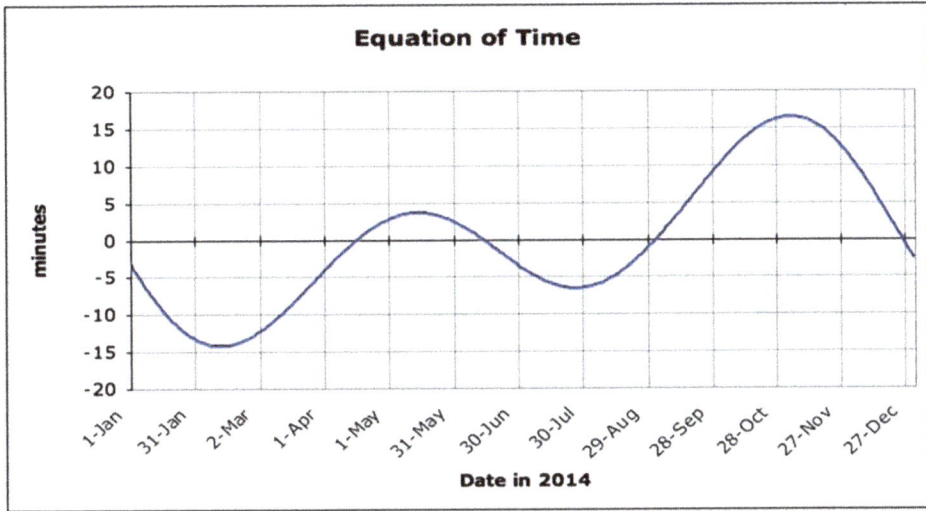

Figure C.1. The equation of time. Credit: Astronomical Applications Department, U.S. Naval Observatory, http://aa.usno.navy.mil/faq/docs/eqtime.php.

before the clock reads 12 noon, i.e. the Sun will cross the meridian when the clock reads 11:44.

In summary, to find when the Sun crosses the meridian read the number of minutes from figure C.1:
- if the sign is negative, the number of minutes is added to 12 noon;
- if the number of minutes is positive, it is subtracted from 12 noon.

The curve in figure C.1 crosses the 0 line at four points, meaning that the Sun crosses the meridian at 12:00 mean solar time four times every year.

C.1 Civil time and the 24 h day

As the mean solar time is different for each longitude, a more standardized and uniform *civil time* system became necessary. The mean solar time at the Greenwich Observatory in the UK is referred to as the Greenwich Mean Time (GMT) and was the first international standard time to be adopted. It served as the basis for standard time and time zones used across the globe. The time zones are approximately 15° wide, which is the equivalent of 1 h.[2] The civil time is the same for the entire time zone and coincides with the mean solar time of the nominal center of the zone.

Greenwich (longitude 0°) is the center of time zone 0 which extends from longitudes 7.5° E to 7.5° W. Time zone +1 is centered at longitude 15° E, extending from longitude 7.5° E to 22.5° E. If the time in Greenwich is 06:00 (6 a.m.) then the civil time in zone +1 is 07:00 (7 a.m.). In zone +2 it is 08:00 (8 a.m.), etc. Similarly time zone −1 is centered at longitude 15° W, extending from longitude 7.5° W to 22.5° W. If the time in Greenwich is 06:00 (6 a.m.) then the civil time in zone −1 is 05:00 (5 a.m.).

---

[2] See the discussion of RA in chapter 3.

In zone −2 it is 04:00 (4 a.m.), etc. There are notable irregularities in the boundaries to accommodate for national and provincial borders, as well as zones with fractional time changes (30 min or 15 min).

The *daylight savings* time affects civil time by moving the clocks one hour forward in the spring (daylight savings time). In the autumn the clocks are moved back one hour and this is the *standard time*.

The mean solar time varies within the zone. For each degree west of the center of the zone, the mean time is 4 min behind the civil time of the zone. For each degree east of the center of the zone, the mean solar time is 4 min ahead of the civil time of the zone. For example, if the center of the zone is at longitude 30° E and the civil time of the zone is 10:00 (10 a.m.), then the mean solar time in a location at longitude 31° E is 10:04 and the mean solar time in a location at longitude 29° E is 09:56, assuming of course that the locations are not separated by an irregular boundary. A 24 h interval is the *mean solar day*, and is measured from midnight to midnight. Note that in the common a.m. and p.m. designations, noon is 12 p.m. and midnight is 12 a.m.

At present, the *Universal Coordinated Time* (UTC) is a 0–24 h clock and is the current basis for civil time. It is an extension of the GMT in that it refers to the same meridian (Greenwich) and combines data from several timekeeping centers worldwide. The UTC makes use of timers based on the so-called atomic clocks. Atomic clocks allow time measurement with a precision of one billionth of a second. As a result, minor irregularities in the UTC are corrected by including 'leap' seconds.

The UT1 system tracks the Earth's rotation using signals from powerful radio sources in space (quasars). The difference between UTC and UT1 is less than 1 s. For most practical purposes the difference between various time systems is of little importance. The time of astronomical events (equinoxes, eclipses, Moon's phases, etc) are usually posted in UTC. The conversion methods outlined in the beginning of this chapter apply to UTC as well.

Online access to UTC, UT1 and others is available at http://time.gov/HTML5/ or http://www.usno.navy.mil/USNO/time.

C.2 Sidereal time

For astronomical measurements, a convenient way of marking time is based on successive transits of a star through the local meridian. This time interval is known as the *sidereal day*. The word sidereal derives from the Latin *sidus*, meaning *star*. By convention, the sidereal day is marked by the transits of the March equinox. The length of one sidereal day is approximately 23 h 56 min and 4 s. It should be noted that the rotation of the Earth is slowing down because of the Moon's gravitational interaction. The Earth's rotation is also affected by shifts of material in the Earth's crust and interior. In addition, as the sidereal time is marked by the transit of the March equinox, the sidereal day is also affected by the precession of the equinoxes and small irregularities in the Earth's spin axis[3].

---

[3] See the discussion of precession in chapter 2.

The *local apparent sidereal time* (LAST) equals the hour angle (HA)[4] of the March equinox, corrected for precession and other fluctuations of the spin axis. As the right ascension (RA) of stars is measured relative to the March equinox, all stars that have a RA equal to the LAST cross the local meridian at the same time at that location. For example, both Iota Orionis and Lambda Orionis have RA 5 h 35 m. Both these stars will cross the meridian of a location when the sidereal clock in that location reads 5 h 35 m. In other words, a star will cross the meridian of the observer when the local sidereal time at the observer's location equals the RA of the star. This is the main reason why the RA is measured in units of time rather than in units of angle.

There is a simple relation between LAST, the HA of a star and the RA of a star:

$$HA = LAST - RA.$$

Therefore, if we know the local sidereal time and the RA of a star, we can locate the meridian of the star (the hour circle of the star, defined in chapter 3). If the star's declination is also known, then the position of the star in the sky is completely defined.

Listings of the sidereal time can be downloaded from http://aa.usno.navy.mil/data/docs/siderealtime.php.

An instant reading of the local sidereal time is available on line at http://tycho.usno.navy.mil/sidereal.html.

---

[4] For HA and RA see the discussion of equatorial coordinates in chapter 3.

Appendix D

Star magnitude systems and the distance modulus

The brightness of the stars we see in the sky is determined by the amount of light emitted from the star and the distance of the star. We use the *apparent* and *absolute magnitude* systems to describe how bright the star appears to an observer on Earth and how bright the star actually is. Both systems use a scale that is inverse: the larger the number the dimmer the star. Also, the numbers used in both systems indicate ratios rather than differences. The two magnitudes are related to each other. The relation between the two magnitudes is given by the *distance modulus* and involves the distance of the star, as will be described below.

D.1 The apparent magnitude system

The *apparent magnitude* system refers to the *apparent brightness*, i.e. the brightness as perceived by an observer on Earth. The system was introduced by the Greek astronomer Hipparchos over 2000 ago and in its original form used six magnitudes: 1 through 6. The brightest stars were assigned magnitude 1. The dimmest stars were assigned magnitude 6.

The symbol m is used to indicate the apparent magnitude. The magnitude scale works as follows. If the magnitudes of two stars differ by 1 unit, then the star with the smaller m is 2.51 times brighter than the star with the larger m[1].

For example:
- If star A has $m = 1$ and star B has $m = 2$ then star A appears 2.51 times brighter than star B.
- If star A has $m = 2$ and star B has $m = 3$ then star A appears 2.51 times brighter than star B.

---

[1] The number 2.51 is used here approximately for the fifth root of 100, which is 2.511886...

If the magnitudes of two stars differ by 2 units, then the star with the smaller m is 2.51^2 times (approximately 6.3 times) brighter than the star with the larger m.

For example:

- If star A has $m = 1$ and star B has $m = 3$ then star A appears 2.51^2 (approximately 6.3) times brighter than B.
- If star A has $m = 4$ and star B has $m = 6$ then star A appears 2.51^2 times brighter than B.

More generally, if the magnitude of object A is mA and the magnitude of object B is mB then object A appears $2.51^{(m\text{B}-m\text{A})}$ times brighter than object B.

For example, if mA $= 1$ and mB $= 6$ then star A is brighter than star B by $2.51^{(6-1)} = 2.51^5 = 100$. Therefore, a decrease of magnitude by 5 units means a 100 times increase in brightness.

Before the invention of the telescope, the system was based on observations with the naked eye. The invention of the telescope allowed observation of stars dimmer than 6, therefore magnitudes larger than 6 were introduced. The use of modern instrumentation makes the magnitude scale more precise, so the units are subdivided into decimals. For example, the magnitude of the star Deneb is 1.25. The system also extends to negative magnitudes. For example, the magnitude of the star Sirius is -1.46.

Table D.1 shows the brightness ratio (brightness of larger m to brightness of smaller m) for given differences in magnitude.

An increase of 5 units in m corresponds to a decrease in brightness by a factor of 100. An increase of 10 units in m corresponds to a decrease in brightness by a factor of 10 000.

Table D.2 shows the reduction in brightness as m increases in steps of 5.

The brightest stars in the sky have magnitudes around -1 to 0. A magnitude of 25 is about the limit of what can be detected using telescopes at present. Stars of magnitude 25 appear 10^{10} (ten billion) times dimmer than the brightest stars! The Sun appears as the brightest object in the sky. For the Sun $m = -26.7$.

Magnitudes apply also to objects that do not emit their own light, for example planets and the Moon, although the values may change because of their changing distances from Earth. The full Moon on average reaches $m = -12.7$ and Venus at her brightest can reach $m = -4.8$.

Table D.1. The brightness ratio for given differences in apparent magnitude.

Difference in m	1	2	3	4	5
Brightness ratio	0.40	0.16	0.06	0.03	0.01

Table D.2. The brightness ratio for given differences in apparent magnitude.

Difference in m	5	10	15	20	25
Brightness ratio	10^{-2}	10^{-4}	10^{-6}	10^{-8}	10^{-10}

D.2 Absolute magnitude

The Sun appears brighter than any star in the sky. But in terms of actual light output (so-called *luminosity*)[2], the Sun is an average star. For example, the star Deneb emits 50 000 times more light than the Sun, meaning that if Deneb were in place of the Sun, we would receive 50 000 times more light! The absolute magnitude is a system that compares the actual light output of stars. The basic idea is as follows.

The star Arcturus (apparent magnitude $m = -0.04$) is about 37 light years (ly) away and emits 200 times more light than the Sun. If we could place the Sun at 37 ly away from Earth, it would look about 200 times dimmer than Arcturus. Using the same thought experiment, we can place all the stars at the same distance from Earth and compare their brightness at this equal setting. This method would eliminate the dependence on distance. For the standard distance we choose 10 parsecs (pc)[3].

We define the *absolute magnitude* as the magnitude the star would have if set at a distance of 10 pc from Earth. The absolute magnitude is designated by M. For example, for the Sun $M = 4.9$ which means that if the Sun was set at a distance of 10 pc (33 ly) away from Earth, it would be a star of apparent magnitude 4.9, i.e. barely visible to the naked eye!

As a rule, if the absolute magnitude is smaller (and remember -100 is smaller than -10) than the apparent magnitude, then the star is farther away than 10 pc.

The reverse is also true: if the apparent magnitude is smaller (and remember -100 is smaller than -10) than the absolute magnitude, then the star is closer than 10 pc. There is a direct relation between the numerical difference between the two magnitudes and the distance, which is discussed next.

D.3 The distance modulus

The relation between absolute magnitude M and apparent magnitude m is

$$M = m + 5 - 5\log(d)$$

where d is the distance of the star in parsecs and log is the logarithm to base 10.

For example, Arcturus has $m = -0.04$ and is about 37 ly away, i.e. in parsecs $d = (37/3.3) = 11$ pc.

Therefore, for Arcturus, $M = -0.04 + 5 - 5\log(11) = -0.04 + 5 - 5 \times 1.05 = -0.29$.

Note that $\log(10) = 1$, therefore, if a star is already at a distance of 10 pc, its absolute and apparent magnitudes are the same. In the example above, Arcturus is at a distance of 11 pc, very close to 10 pc, therefore it is not surprising that M and m for Arcturus are about the same. If the star is farther away than 10 pc, the absolute magnitude is a smaller number than the apparent magnitude. Remember that -5 is smaller than -4 and that smaller in the magnitude systems means brighter.

---

[2] See the discussion of luminosity in chapter 8.
[3] One parsec equals about 3.3 ly. See the discussion on the small angle formula in appendix B.

For example Canopus, the second brightest star in the sky, has apparent magnitude $m = -0.72$ and distance 310 ly (310/3.3 = 94 pc). The absolute magnitude of Canopus is:

$$M = m + 5 - 5\log(d) = -0.72 + 5 - 5 \times \log(94) = -0.72 + 5 - 5 \times 2 = -5.7.$$

In other words, if Canopus was 10 pc away it would outshine Venus at her best (for Venus $m = -4.8$).

On the other hand Sirius, the brightest star in the sky, has $m = -1.46$ and a distance of 8.6 ly (8.6/3.3 = 2.6 pc). The absolute magnitude of Sirius is:

$$M = m + 5 - 5\log(d) = -1.46 + 5 - 5 \times \log(2.6) = -1.46 + 5 - 5 \times 0.42 = 1.44.$$

Thus if both Sirius and Canopus were placed at 10 pc from Earth, Canopus would appear $2.5^{(5.7+1.44)}$, over 700 times brighter than Sirius!

The difference $m-M$ is the *distance modulus*, because it is a direct measure of the distance of the star. This can be seen by rearranging the relation between m, M and d above. From $M = m + 5 - 5\log(d)$ it follows that:

$$\log(d) = 1 + (m - M)/5.$$

Therefore, the larger the distance modulus $(m - M)$ the larger the distance is. If $m = M$ the star is at 10 pc, because $\log(10) = 1$.

We can extract the distance from the logarithm, recalling that $a = \log(x)$ means that $x = 10^a$. For example if $\log(x) = 3$, then $x = 10^3$. From $\log(d) = 1 + (m - M)/5$ we find the formula for the distance in terms of m and M:

$$d = 10^{1 + (m-M)/5}.$$

We can use this formula directly to calculate the distance of a star in parsecs, once the apparent and absolute magnitude are known. For example, Polaris has $m = 1.96$ and $M = -3.65$; therefore

$$d = 10^{1+(1.96+3.65)/5} \quad \text{or} \quad d = 10^{1+1.122} = 10^{2.122} = 132 \text{ pc}.$$

The distance formula is very useful, because we can always find m by directly observing the star, and we can 'guess' M from the HR diagram. As discussed in chapter 8, the spectral type of the star and the luminosity class allow us to approximately locate the star in the HR diagram. From the point representing the star in the HR diagram we can read on the vertical axis the luminosity of the star in units of the Sun's luminosity. From the luminosity we can calculate M. Inserting M and m in the above equation yields the distance of the star in parsecs. This method of determining the distance is known as *spectroscopic parallax*, although there is no angle involved in the procedure.

IOP Concise Physics

Visual Astronomy
A guide to understanding the night sky
Panos Photinos

Appendix E

Bibliography

A detailed and easy to follow stargazing guide for the entire sky can be found in:

Moore P 2001 *Stargazing: Astronomy without a Telescope* 2nd edn (Cambridge: Cambridge University Press)

Constellation maps, figures, coordinates and transit times and other details useful in observational work are contained in:

Bakich M E 1995 *The Cambridge Guide to the Constellations* (Cambridge: Cambridge University Press)

More advanced discussion of coordinate systems and spherical trigonometry equations can be found in:

Roy A E and Clarke D 2003 *Astronomy: Principles and Practice* 4th edn (London: Taylor and Francis)

Birney D S, Gonzalez G and Oesper D 2006 *Observational Astronomy* (Cambridge: Cambridge University Press)

Introduction to the celestial sphere, Kepler's laws, the solar system, the physical properties of planets and stars, in a very accessible level of discussion can be found in:

Chaisson E and McMillan S 2013 *Astronomy Today* 8th edn (Reading, MA: Addison-Wesley)

Koupelis T 2014 *In Quest of the Universe* 7th edn (Jones and Bartlett)

Arny T T and Schneider S E 2013 *Explorations: Introduction to Astronomy* 7th edn (New York: McGraw Hill)

For a more advanced introduction, see:

Roy A E and Clarke D 2003 *Astronomy: Principles and Practice* 4th edn (London: Taylor and Francis)

Holliday K 1999 *Introductory Astronomy* (John Wiley & Sons)

Optics of telescopes and imaging, as well as an introduction to methods of analysis, in a level suitable for students intending to pursue a major in astronomy are discussed in:

Kitchin C R 2013 *Telescopes and Techniques* 3rd edn (Berlin: Springer)

Birney D S, Gonzalez G and Oesper D 2006 *Observational Astronomy* (Cambridge: Cambridge University Press)

An interesting and enjoyable account of the development of calendars and measuring time is given in:

Steel D 1999 *Marking Time: The Epic Quest to Invent the Perfect Calendar* (John Wiley & Sons)

Two classic works of popularized astronomy:

Rudaux L and De Vaucouleurs G 1962 *Larousse Encyclopedia of Astronomy* (Prometheus Press)

Flammarion C 1964 *The Flammarion Book of Astronomy* (Simon & Schuster)

A scholarly work on the origin and meaning of star names is the re-edition of the classic work:

Allen R H 1963 *Star-names and Their Lore and Meaning* Dover Edition

www.ingramcontent.com/pod-product-compliance
Lightning Source LLC
Chambersburg PA
CBHW080426270326

41929CB00018B/3185